Annals of Mathematics Studies

Number 32

ANNALS OF MATHEMATICS STUDIES

Edited by Emil Artin and Marston Morse

1. Algebraic Theory of Numbers, *by* HERMANN WEYL
3. Consistency of the Continuum Hypothesis, *by* KURT GÖDEL
6. The Calculi of Lambda-Conversion, *by* ALONZO CHURCH
7. Finite Dimensional Vector Spaces, *by* PAUL R. HALMOS
10. Topics in Topology, *by* SOLOMON LEFSCHETZ
11. Introduction to Nonlinear Mechanics, *by* N. KRYLOFF *and* N. BOGOLIUBOFF
14. Lectures on Differential Equations, *by* SOLOMON LEFSCHETZ
15. Topological Methods in the Theory of Functions of a Complex Variable, *by* MARSTON MORSE
16. Transcendental Numbers, *by* CARL LUDWIG SIEGEL
17. Problème Général de la Stabilité du Mouvement, *by* M. A. LIAPOUNOFF
19. Fourier Transforms, *by* S. BOCHNER *and* K. CHANDRASEKHARAN
20. Contributions to the Theory of Nonlinear Oscillations, Vol. I, *edited by* S. LEFSCHETZ
21. Functional Operators, Vol. I, *by* JOHN VON NEUMANN
22. Functional Operators, Vol. II, *by* JOHN VON NEUMANN
23. Existence Theorems in Partial Differential Equations, *by* DOROTHY L. BERNSTEIN
24. Contributions to the Theory of Games, Vol. I, *edited by* H. W KUHN *and* A. W. TUCKER
25. Contributions to Fourier Analysis, *by* A. ZYGMUND, W. TRANSUE, M. MORSE, A. P. CALDERON, *and* S. BOCHNER
26. A Theory of Cross-Spaces, *by* ROBERT SCHATTEN
27. Isoperimetric Inequalities in Mathematical Physics, *by* G. POLYA *and* G. SZEGO
28. Contributions to the Theory of Games, Vol. II, *edited by* H. KUHN *and* A. W. TUCKER
29. Contributions to the Theory of Nonlinear Oscillations, Vol. II, *edited by* S. LEFSCHETZ
30. Contributions to the Theory of Riemann Surfaces, *edited by* L. AHLFORS *et al.*
31. Order-Preserving Maps and Integration Processes, *by* EDWARD J. MCSHANE
32. Curvature and Betti Numbers, *by* K. YANO *and* S. BOCHNER
33. Contributions to the Theory of Partial Differential Equations, *edited by* L. BERS, S. BOCHNER, *and* F. JOHN

CURVATURE
AND
BETTI NUMBERS

By K. Yano and S. Bochner

Princeton, New Jersey
Princeton University Press
1953

London: Geoffrey Cumberlege, Oxford University Press
L. C. Card 53-6583

Printed in the United States of America

PREFACE

This tract gives a first systematic account of a topic in differential geometry in the large, that is, of a topic on curvature and Betti numbers, recently inaugurated by Professor S. Bochner.

In the hope that the tract might also be of use as a survey of present-day differential geometry in several of its aspects, and also in order to fix our notation, all pre-requisites from differential geometry as such are presented in Chapter 1, virtually independently.

The other Chapters contain the recent work of Professor Bochner on differential geometry in the large, and other results closely related to it.

These Chapters contain only a part of recent work in this field, but very fortunately for myself and also for the reader, Professor Bochner was kind enough to add a chapter of supplements and from which the reader will learn wider aspects of this very interesting topic.

I wish to express here my hearty thanks to Professor O. Veblen who gave me the opportunity to stay at the Institute for Advanced Study and also to Professor D. Montgomery who gave me a chance to give a lecture in his seminar on which the first draft of the tract was based.

Professor Bochner not only added the last Chapter which is the most important and the most interesting part of the book, but also gave me many valuable suggestions on the first eight Chapters. It is my pleasant duty to express here my sincere gratitude to Professor Bochner without whose kindness this book would not be possible.

Kentaro Yano

Institute for Advanced Study
May 5, 1952

CONTENTS

CHAPTER I

RIEMANNIAN MANIFOLD

1. RIEMANNIAN MANIFOLD

We take a Hausdorff space with a given system of neighborhoods $\{U\}$, such that each neighborhood U can be put in one-to-one reciprocal continuous correspondence with the interior of a hypersphere

$$\sum_{i=1}^{n} (x^i - x_0^i)^2 = r^2$$

in an n-dimensional Euclidean space, and such a space we will call an n-dimensional manifold, where, and in the following, Roman indices run over the values $1, 2, \ldots, n$.

This correspondence between points in a neighborhood of the manifold and points in the inside of a hypersphere is called a coordinate system, and the coordinates (x^i) of the point in the Euclidean space which corresponds to the point P in the manifold are called coordinates of the point P in this coordinate system. Moreover, a neighborhood endowed with a coordinate system is called a coordinate neighborhood.

If, in a neighborhood U, two coordinate systems

$$(x^1, x^2, \ldots, x^n) \quad \text{and} \quad (x'^1, x'^2, \ldots, x'^n)$$

are given, then there is a one-to-one reciprocal continuous correspondence between these two coordinate systems which can be expressed by the equations

$$x^i = x^i(x'^1, x'^2, \ldots, x'^n) \tag{1.1}$$

or inversely

$$x'^i = x'^i(x^1, x^2, \ldots, x^n) \tag{1.2}$$

Equations (1.1) or (1.2) define a so-called coordinate transformation.

If the functions $x^i(x'^a)$ and $x'^i(x^a)$ are of class C^r, that is to say, if they admit continuous partial derivatives of the first, the second, ..., the r-th derivatives, and if, when $r \geq 1$, the Jacobians

$$\left|\frac{\partial x^i}{\partial x'^a}\right| \quad \text{and} \quad \left|\frac{\partial x'^a}{\partial x^i}\right|$$

are different from zero for any coordinate transformation in the manifold, we say that the manifold is of class C^r.

It is evident that, in an n-dimensional manifold of class C^r, if we have functions $f^i(x^a)$ satisfying the above mentioned conditions, where (x^a) is an original coordinate system in a neighborhood U, then, on putting

$$x'^i = f^i(x^a)$$

we can introduce (x'^i) as a new coordinate system in U. We shall call such a coordinate system an allowable coordinate system in U.

Now, if the manifold can be covered entirely by a finite number of neighborhoods $U_1, U_2, \ldots, U_N$, then the manifold is said to be compact. As a rule, our manifolds will be compact.

We assume sometimes also the orientability of the manifold. If (x^i) and (x'^i) are two allowable coordinate systems in a coordinate neighborhood U, then the Jacobian

$$\left|\frac{\partial x^i}{\partial x'^a}\right| = J$$

is different from zero throughout the coordinate neighborhood U, and consequently, being a continuous function of the point in U, it has the same sign throughout U. If this sign is positive, we say that these coordinate systems are positively related, and if it is negative, we say that they are negatively related.

If there exists a subset of the set of all allowable coordinate neighborhoods such that it covers the whole manifold and any coordinate systems which belong to this subset and which are valid in the same neighborhood are positively related, then we say that the manifold is orientable.

We now assume that, with each coordinate neighborhood U in our n-dimensional manifold of class C^r, there is associated a positive definite quadratic differential form in the differentials dx^i,

$$ds^2 = g_{jk}dx^j dx^k \qquad (g_{jk} = g_{kj}) \tag{1.3}$$

which does not depend on the coordinate system used, where the coefficients $g_{jk}(x)$ are functions of coordinates $(x^1, x^2, \ldots, x^n)$ of class C^{r-1}, and repeated indices represent the summation over their range.

Geometrically, we interpret (1.3) as a formula which gives the infinitesimal distance ds between two points (x^i) and $(x^i + dx^i)$, the length of a curve $x^i = x^i(t)$ $(t_1 \leq t \leq t_2)$ being given by

$$s = \int_{t_1}^{t_2} \sqrt{g_{jk} \frac{dx^j}{dt} \frac{dx^k}{dt}}\, dt \tag{1.4}$$

and we call (1.3) the fundamental metric form of the manifold.

An n-dimensional manifold of class C^r in which a fundamental metric form (1.3) is given is called an n-dimensional Riemannian manifold of class C^r and the theory of such manifolds is called Riemannian geometry. (L. P. Eisenhart [1]).

2. TENSOR ALGEBRA

If in a coordinate system (x^i) we are given the form (1.3) and if in another coordinate system we correspondingly put

$$ds'^2 = g'_{jk}(x')dx'^j dx'^k$$

then we must have

$$ds' = ds$$

In general, if an object is represented by f in a coordinate system (x^i) and by f' in any other coordinate system (x'^i), and if we have

$$f' = f \tag{1.5}$$

then we call this object a scalar and f and f' its components in the respective coordinate systems (x^i) and (x'^i). Thus, ds is the component of a scalar.

Next from (1.2), we have

$$dx'^i = \frac{\partial x'^i}{\partial x^r} dx^r$$

In general, if an object is represented by n quantities v^i in a coordinate system (x^i) and by v'^i in any other coordinate system (x'^i), and if we have

$$v'^i = \frac{\partial x'^i}{\partial x^r} v^r \tag{1.6}$$

then we call this object a contravariant vector and v^i and v'^i its components in respective coordinate systems (x^i) and (x'^i). Thus, dx^i are components of a contravariant vector in the coordinate system (x^i).

If an object is defined at every point of a coordinate neighborhood U, then its components are functions of (x^i). We call such an object a field.

If we denote by $f(x)$ and $f'(x')$ the components of a scalar field in respective coordinate systems (x^i) and (x'^i), then we have

$$f'(x') = f(x)$$

from which, by partial differentiation,

$$\frac{\partial f'}{\partial x'^j} = \frac{\partial x^s}{\partial x'^j} \frac{\partial f}{\partial x^s}$$

In general, if an object is represented by n quantities v_j in a coordinate system (x^i) and by v'_j in any other coordinate system (x'^i), and if we have

$$v'_j = \frac{\partial x^s}{\partial x'^j} v_s \tag{1.7}$$

then we call this object a covariant vector and v_j and v'_j its components in respective coordinate systems (x^i) and (x'^i). If $f(x)$ is a component of a scalar field, then $\partial f/\partial x^i$ are components of a covariant vector. We call such a special covariant vector the gradient of the scalar field f.

From the assumption $ds' = ds$, we have

$$g'_{jk} = \frac{\partial x^s}{\partial x'^j} \frac{\partial x^t}{\partial x'^k} g_{st}$$

In general, if an object is represented by n^{p+q} quantities

$$T^{i_1 i_2 \cdots i_p}{}_{j_1 j_2 \cdots j_q}$$

in a coordinate system (x^i) and by

$$T'^{i_1 i_2 \cdots i_p}{}_{j_1 j_2 \cdots j_q}$$

in any other coordinate system (x'^i), and if we have

$$T'^{i_1 i_2 \cdots i_p}{}_{j_1 j_2 \cdots j_q} = \tag{1.8}$$

$$\frac{\partial x'^{i_1}}{\partial x^{r_1}} \frac{\partial x'^{i_2}}{\partial x^{r_2}} \cdots \frac{\partial x'^{i_p}}{\partial x^{r_p}} \frac{\partial x^{s_1}}{\partial x'^{j_1}} \frac{\partial x^{s_2}}{\partial x'^{j_2}} \cdots \frac{\partial x^{s_q}}{\partial x'^{j_q}} T^{r_1 r_2 \cdots r_p}{}_{s_1 s_2 \cdots s_q}$$

then we call this object a mixed tensor of contravariant valency p and of covariant valency q, and

$$T^{i_1 i_2 \cdots i_p}{}_{j_1 j_2 \cdots j_q}$$

and

$$T'^{i_1 i_2 \cdots i_p}{}_{j_1 j_2 \cdots j_q}$$

its components in respective coordinate systems (x^i) and (x'^i). A tensor having only contravariant valency is called a contravariant tensor and a tensor having only covariant valency a covariant tensor. Thus, $g_{jk}(x)$ are components of a covariant tensor field.

Since we have assumed that (1.3) is positive definite, we have

$$g \equiv \begin{vmatrix} g_{11} & g_{12} & \cdots & g_{1n} \\ g_{21} & g_{22} & \cdots & g_{2n} \\ & \cdots & & \\ & \cdots & & \\ g_{n1} & g_{n2} & \cdots & g_{nn} \end{vmatrix} > 0 \tag{1.9}$$

and consequently we can define g^{ij} by

$$g^{ij} = \frac{(\text{the cofactor of } g_{ji} \text{ in } g)}{g} \tag{1.10}$$

Thus we have

$$(1.11) \qquad g^{ij}g_{jk} = \delta^i_k = \begin{cases} 1 & \text{for } i = k \\ 0 & \text{for } i \neq k \end{cases}$$

and the δ^i_k here defined is known as Kronecker's delta.

It will be easily seen that $g^{ij} = g^{ji}$ are components of a contravariant tensor, and δ^i_k are components of a mixed tensor. We call g_{jk}, g^{ij} and δ^i_k fundamental covariant, contravariant and mixed tensors respectively.

If, for instance, components $T^i{}_{jk}$ of a tensor satisfy

$$T^i{}_{jk} = T^i{}_{kj}$$

we say that they are symmetric in j and k, and if they satisfy

$$T^i{}_{jk} = - T^i{}_{kj}$$

we say that they are anti-symmetric in j and k.

It is easily proved that if the components of a tensor are symmetric or anti-symmetric in a coordinate system, then they are so in any other coordinate system. If the components of a contravariant or covariant tensor are symmetric (anti-symmetric) in all the indices, then we call the tensor a symmetric (anti-symmetric) tensor. The g_{jk} and g^{ij} are both symmetric tensors.

We shall next state some algebraic operations which can be applied to tensors.

(i) Addition and subtraction.

Let, for instance, $R^i{}_{jk}$ and $S^i{}_{jk}$ be components of two tensors of the same type, then

$$R^i{}_{jk} + S^i{}_{jk} = T^i{}_{jk}$$

are components of a tensor of the same type which is called the sum of two given tensors. The difference of two tensors is defined in an analogous way.

(ii) Multiplication.

Let, for instance, $R^i{}_j$ and S_{kl} be components of two tensors of any type, then

$$R^i{}_jS_{kl} = T^i{}_{jkl}$$

are components of a tensor of the type indicated by the position of the indices, and it is called the product of the two given tensors.

(iii) Contraction.

Let, for instance, $T^i{}_{jkl}$ be the components of a mixed tensor. The quantities

$$T^s{}_{jks} = T_{jk}$$

are components of a tensor having two less indices than the original one, and in this case, we say that we have contracted $T^i{}_{jkl}$ with respect to i and l, obtaining T_{jk}.

(iv) Raising and lowering of indices.

If λ^i are components of a contravariant vector, then $g_{jk}\lambda^i$ are, by (ii), components of a mixed tensor, and consequently $g_{jk}\lambda^k$ are, by (iii), components of a covariant vector. We denote it by $\lambda_j = g_{jk}\lambda^k$. Similarly if μ_k are components of a covariant vector, then $g^{ij}\mu_k$ are, by (ii), components of a mixed tensor and consequently, $g^{ij}\mu_j$ are, by (iii), components of a contravariant vector. We denote it by $\mu^i = g^{ij}\mu_j$. It is evident that if

$$\mu_j = \lambda_j = g_{jk}\lambda^k$$

then

$$\mu^i = \lambda^i$$

We will say that λ^i and λ_i are conjugate to one another, and we are introducing an object which can be represented by λ^i and λ_i alternatively. We call it a vector, and λ^i its contravariant components, and λ_i its covariant components.

The same thing can be stated for the components of a tensor as is shown in the following examples:

$$T^i{}_{jk} \longrightarrow T_{ijk} = g_{is}T^s{}_{jk}$$

$$T_{ijk} \longrightarrow T_{ij}{}^k = T_{ijs}g^{sk}$$

In the first example, we say that we lowered the index i, and in the second, we say that we raised the index k, and that we are dealing with components of the same tensor.

(v) Symmetrization and anti-symmetrization.

Consider, for example, a covariant tensor T_{ijk}, and form the sum of all the components obtainable from T_{ijk} by taking all the possible permutations of the indices i, j, and k, and divide it by $3!$ (number of the all possible permutations). We denote the resulting object by

$$T_{(ijk)} = \frac{1}{3!}(T_{ijk} + T_{jki} + T_{kij} + T_{jik} + T_{kji} + T_{ikj})$$

and call it the symmetric part of T_{ijk}. It is easily shown that $T_{(ijk)}$ are components of a symmetric covariant tensor, and the operation $T_{ijk} \to T_{(ijk)}$ is called symmetrization of T_{ijk}. If the original tensor is symmetric, then we have $T_{(ijk)} = T_{ijk}$.

Consider again, for example, a covariant tensor T_{ijk}, and all components obtainable from T_{ijk} by all possible permutations. Next, give a plus sign to a component obtained from T_{ijk} by an even permutation and a minus sign to a component obtained from T_{ijk} by an odd permutation, and form the algebraic sum of these components, and divide it by $3!$. We denote the resulting object by

$$T_{[ijk]} = \frac{1}{3!}(T_{ijk} + T_{jki} + T_{kij} - T_{jik} - T_{kji} - T_{ikj})$$

and call it the anti-symmetric part of T_{ijk}. It is easily shown that $T_{[ijk]}$ are components of an anti-symmetric covariant tensor, and the operation $T_{ijk} \to T_{[ijk]}$ is called anti-symmetrization of T_{ijk}. If the original tensor is anti-symmetric, then we have $T_{[ijk]} = T_{ijk}$.

Now, the formula (1.3) shows that the length of the contravariant vector dx^i is ds, and similarly we define the length λ of a contravariant vector λ^i by

$$(\lambda)^2 = g_{jk}\lambda^j\lambda^k \tag{1.12}$$

If we denote the covariant components of this vector by λ_i, then the above formula may be written in the following various forms:

$$(\lambda)^2 = g_{jk}\lambda^j\lambda^k = \lambda_k\lambda^k = \lambda^j\lambda_j = g^{jk}\lambda_j\lambda_k$$

A vector whose length is unity is called a unit vector. If λ^i and μ^i are both unit vectors, then we have

$$g_{jk}\lambda^j\lambda^k = 1 \qquad g_{jk}\mu^j\mu^k = 1$$

and consequently we can prove that

$$(g_{jk}\lambda^j\mu^k)^2 \leq 1$$

Thus, we define the angle θ between two unit vectors λ^i and μ^i by

$$\cos\theta = g_{jk}\lambda^j\mu^k \tag{1.13}$$

and the angle θ between two arbitrary vectors u^i and v^i is given by

$$\cos\theta = \frac{g_{jk}u^j v^k}{u\ v} \tag{1.14}$$

Equation (1.14) gives

$$u\ v\cos\theta = g_{jk}u^jv^k = u_kv^k = u^j\ v_j = g^{jk}u_jv_k$$

and this is called the inner product of two vectors u^i and v^i.

From (1.14), we see that two vectors u^i and v^i are orthogonal to each other if

$$g_{jk}u^jv^k = 0$$

Next, from the transformation law of g_{jk}:

$$g'_{jk} = \frac{\partial x^s}{\partial x'^j}\frac{\partial x^t}{\partial x'^k}g_{st}$$

we find

$$\sqrt{g'} = \left|\frac{\partial x}{\partial x'}\right|\sqrt{g}$$

and, on the other hand, the transformation law of $dx^1dx^2 \ldots dx^n$ in an n-tuple integral is

$$dx'^1dx'^2 \ldots dx'^n = \left|\frac{\partial x'}{\partial x}\right| dx^1dx^2 \ldots dx^n$$

Thus, from these two equations, we get

$$\sqrt{g'}\, dx'^1dx'^2 \ldots dx'^n = \sqrt{g}\, dx^1dx^2 \ldots dx^n$$

which shows that

$$dv = \sqrt{g}\, dx^1 dx^2 \dots dx^n \tag{1.15}$$

is a scalar. We define the volume element of our Riemannian manifold to be dv .

3. TENSOR CALCULUS

Take a curve $x^i(t)$ joining two points $P_1(x^i(t_1))$ and $P_2(x^i(t_2))$ and introduce its length

$$I = \int_{t_1}^{t_2} \sqrt{g_{jk} \frac{dx^j}{dt} \frac{dx^k}{dt}}\, dt$$

If another curve

$$\bar{x}^i(t) = x^i(t) + \epsilon u^i(t) \qquad (\epsilon\text{: infinitesimal})$$

passes through P_1 and P_2 (and consequently $u^i(t_1) = u^i(t_2) = 0$) and is infinitesimally close to $x^i(t)$, and if we denote by δI the first variation of the length integral I , then δI is given by

$$\delta I = \epsilon \int_{t_1}^{t_2} \left[\frac{\partial F}{\partial x^i} - \frac{d}{dt}\left(\frac{\partial F}{\partial \dot{x}^i}\right) \right] u^i\, dt$$

where we have put

$$F = \sqrt{g_{jk}(x)\dot{x}^j\dot{x}^k} \qquad \text{and} \qquad \dot{x}^i = \frac{dx^i}{dt}$$

We call the curve for which $\delta I = 0$ for any u^i a geodesic in our Riemannian manifold.

A geodesic must satisfy the so-called Euler differential equations

$$\lambda_i \equiv \frac{d}{dt}\left(\frac{\partial F}{\partial \dot{x}^i}\right) - \frac{\partial F}{\partial x^i} = 0$$

and it will be proved that λ_i are covariant components of a vector.

Taking the arc length s as parameter along the geodesic, and calculating the contravariant components λ^i of λ_i , we find

(1.16) $$\lambda^i \equiv \frac{d^2x^i}{ds^2} + \left\{{i \atop jk}\right\} \frac{dx^j}{ds}\frac{dx^k}{ds} = 0$$

where $\left\{{i \atop jk}\right\}$ are defined by

(1.17) $$\left\{{i \atop jk}\right\} = \frac{1}{2} g^{is} \left(\frac{\partial g_{sj}}{\partial x^k} + \frac{\partial g_{sk}}{\partial x^j} - \frac{\partial g_{jk}}{\partial x^s} \right)$$

and are called Christoffel symbols.

It will be easily verified that the Christoffel symbols satisfy the following identities:

(1.18) $$\frac{\partial g_{jk}}{\partial x^l} - g_{sk} \left\{{s \atop jl}\right\} - g_{js} \left\{{s \atop kl}\right\} = 0$$

(1.19) $$\frac{\partial g^{ij}}{\partial x^k} + g^{sj} \left\{{i \atop sk}\right\} + g^{is} \left\{{j \atop sk}\right\} = 0$$

(1.20) $$\left\{{s \atop js}\right\} = \frac{1}{\sqrt{g}} \frac{\partial \sqrt{g}}{\partial x^j} = \frac{\partial \log \sqrt{g}}{\partial x^j}$$

Now, from the fact that λ^i in (1.16) are contravariant components of a vector, we can find the following transformation law of the Christoffel symbols under a coordinate transformation:

(1.21) $$\frac{\partial^2 x^r}{\partial x'^j \partial x'^k} = \frac{\partial x^r}{\partial x'^i} \left\{{i \atop jk}\right\}' - \frac{\partial x^s}{\partial x'^j}\frac{\partial x^t}{\partial x'^k} \left\{{r \atop st}\right\}$$

(1.22) $$\frac{\partial^2 x'^i}{\partial x^s \partial x^t} = \frac{\partial x'^i}{\partial x^r} \left\{{r \atop st}\right\} - \frac{\partial x'^j}{\partial x^s}\frac{\partial x'^k}{\partial x^t} \left\{{i \atop jk}\right\}'$$

If $f(x)$ is the component of a scalar field, then it is evident that df is also the component of a scalar, and that $\partial f/\partial x^i$ are components of a covariant vector. We call df the covariant differential of the scalar f and $\partial f/\partial x^i$ the covariant derivative of the scalar f and denote them respectively by

(1.23) $$\delta f = df$$

(1.24) $$f_{;j} = \frac{\partial f}{\partial x^j}$$

If $v^i(x)$ are components of a contravariant vector field, then

dv^i are not necessarily components of a contravariant vector. But, combining the transformation law of dv^i with that of $\{{i \atop jk}\}$, we can prove that

(1.25) $$\delta v^i = dv^i + v^j \{{i \atop jk}\} dx^k$$

are components of a contravariant vector and

(1.26) $$v^i{}_{;k} = \frac{\partial v^i}{\partial x^k} + v^j \{{i \atop jk}\}$$

are components of a mixed tensor. We call δv^i covariant differential of v^i and $v^i{}_{;k}$ covariant derivative of v^i .

Similarly, if $v_j(x)$ are components of a covariant vector field, then dv_j are not necessarily components of a covariant vector, but we can prove that

(1.27) $$\delta v_j = dv_j - v_i \{{i \atop jk}\} dx^k$$

are components of a covariant vector and

(1.28) $$v_{j;k} = \frac{\partial v_j}{\partial x^k} - v_i \{{i \atop jk}\}$$

are components of a covariant tensor. We call δv_j covariant differential of v_j and $v_{j;k}$ covariant derivative of v_j .

This operation of covariant differentiation may be applied to a general tensor, say, to $T^i{}_{jk}$:

(1.29) $$\delta T^i{}_{jk} = dT^i{}_{jk} + T^s{}_{jk}\{{i \atop sl}\}dx^l - T^i{}_{sk}\{{s \atop jl}\}dx^l - T^i{}_{js}\{{s \atop kl}\}dx^l$$

(1.30) $$T^i{}_{jk;l} = \frac{\partial T^i{}_{jk}}{\partial x^l} + T^s{}_{jk}\{{i \atop sl}\} - T^i{}_{sk}\{{s \atop jl}\} - T^i{}_{js}\{{s \atop kl}\}$$

We call $\delta T^i{}_{jk}$, which is a tensor of the same type as $T^i{}_{jk}$, the covariant differential of $T^i{}_{jk}$ and $T^i{}_{jk;l}$, which is a tensor having one more covariant index than $T^i{}_{jk}$, the covariant derivative of $T^i{}_{jk}$.

If we apply this operation of covariant differentiation to the tensors g_{jk} , g^{ij} , and δ^i_k , we get

(1.31) $$g_{jk;l} = \frac{\partial g_{jk}}{\partial x^l} - g_{sk} \{{s \atop jl}\} - g_{js} \{{s \atop kl}\} = 0$$

(1.32) $$g^{ij}{}_{;k} = \frac{\partial g^{ij}}{\partial x^k} + g^{sj}\{{i \atop sk}\} + g^{is}\{{j \atop sk}\} = 0$$

(1.33) $$\delta^i_{j;k} = \frac{\partial \delta^i_j}{\partial x^k} + \delta^s_j\{{i \atop sk}\} - \delta^i_s\{{s \atop jk}\} = 0$$

Thus, the tensors g_{jk} , g^{ij} , and δ^i_k are all constant under covariant differentiation.

It will be easily verified that the covariant differentiation obeys the rules of ordinary differentiation:

$$\delta(R^i{}_{jk} \pm S^i{}_{jk}) = \delta R^i{}_{jk} \pm \delta S^i{}_{jk}$$

$$\delta(R^i{}_j S_{kl}) = (\delta R^i{}_j)S_{kl} + R^i{}_j(\delta S_{kl})$$

and

$$(R^i{}_{jk} \pm S^i{}_{jk})_{;l} = R^i{}_{jk;l} \pm S^i{}_{jk;l}$$

$$(R^i{}_j S_{kl})_{;m} = (R^i{}_{j;m})S_{kl} + R^i{}_j(S_{kl;m})$$

If we are given a covariant vector field $v_j(x)$, then we can form an anti-symmetric tensor

(1.34) $$v_{j;k} - v_{k;j} = \frac{\partial v_j}{\partial x^k} - \frac{\partial v_k}{\partial x^j}$$

which is independent of the Christoffel symbols. It is called the curl of the covariant vector v_j .

Similarly, if we are given an anti-symmetric tensor field $\xi_{i_1 i_2 \cdots i_p}$, then we can form an anti-symmetric tensor

(1.35) $$(n + 1)\xi_{[i_1 i_2 \cdots i_p;j]}$$

$$= \frac{\partial \xi_{i_1 i_2 \cdots i_p}}{\partial x^j} - \frac{\partial \xi_{j i_2 \cdots i_p}}{\partial x^{i_1}} - \frac{\partial \xi_{i_1 j i_3 \cdots i_p}}{\partial x^{i_2}} - \cdots - \frac{\partial \xi_{i_1 i_2 \cdots i_{p-1} j}}{\partial x^{i_p}}$$

which is independent of the Christoffel symbols. It is called the curl of the anti-symmetric covariant tensor

$$\xi_{i_1 i_2 \cdots i_p}$$

If we are given a contravariant vector field $v^i(x)$, then we can form a scalar

$$(1.36) \qquad v^i{}_{;i} = \frac{\partial v^i}{\partial x^i} + v^j \left\{ {i \atop ji} \right\} = \frac{1}{\sqrt{g}} \frac{\partial \sqrt{g}\, v^i}{\partial x^i}$$

which depends only on $\sqrt{g}$. This is called the divergence of the contravariant vector v^i .

The divergence of a covariant vector v_j is defined as the scalar

$$(1.37) \qquad g^{jk}(v_{j;k})$$

and the divergence of a covariant tensor $\xi_{i_1 i_2 \dots i_p}$ as the tensor

$$(1.38) \qquad g^{ij}\, \xi_{i i_2 \dots i_p ; j}$$

Now, if we are given a scalar field $f(x)$, then we can form its gradient $f_{;j}$ and the square of its length:

$$(1.39) \qquad \Delta_1 f = g^{ij} f_{;i} f_{;j}$$

This is called Beltrami's differential parameter of the first kind of the scalar field $f(x)$.

With the gradient $f_{;i}$, we can form its divergence:

$$(1.40) \qquad \Delta_2 f = g^{ij}(f_{;i})_{;j} = (g^{ij} f_{;i})_{;j}$$

This is called Beltrami's differential parameter of the second kind of the scalar field $f(x)$. It is also called the Laplacean of $f(x)$ and is denoted by

$$(1.41) \qquad \Delta f = g^{ij} f_{;i;j}$$

4. CURVATURE TENSORS

For a scalar $f(x)$, the covariant derivative of $f(x)$ is given by

$$f_{;j} = \frac{\partial f}{\partial x^j}$$

and the second covariant derivative is given by

$$f_{;j;k} = \frac{\partial^2 f}{\partial x^j \partial x^k} - \frac{\partial f}{\partial x^i}\left\{{i \atop jk}\right\}$$

Thus, we see that

$$f_{;j;k} - f_{;k;j} = 0$$

However, for vectors and tensors, successive covariant differentiations are not commutative in general. Thus, for example, for a contravariant vector v^i, we obtain

(1.43) $$v^i{}_{;k;l} - v^i{}_{;l;k} = v^j R^i{}_{jkl}$$

where[1]

(1.44) $$R^i{}_{jkl} = \frac{\partial\left\{{i \atop jk}\right\}}{\partial x^l} - \frac{\partial\left\{{i \atop jl}\right\}}{\partial x^k} + \left\{{s \atop jk}\right\}\left\{{i \atop sl}\right\} - \left\{{s \atop jl}\right\}\left\{{i \atop sk}\right\}$$

are components of a mixed tensor called Riemann-Christoffel curvature tensor,[1] and this tensor needs not be zero.

Similarly, if we take a covariant vector v_j, then we have

(1.45) $$v_{j;k;l} - v_{j;l;k} = -v_i R^i{}_{jkl}$$

and if we take a general tensor $T^i{}_{jk}$, for example, then we have

(1.46) $$T^i{}_{jk;l;m} - T^i{}_{jk;m;l} = T^s{}_{jk}R^i{}_{slm} - T^i{}_{sk}R^s{}_{jlm} - T^i{}_{js}R^s{}_{klm}$$

Formulas (1.43), (1.45), and (1.46) are called the Ricci formulas.

From the curvature tensor $R^i{}_{jkl}$, we get, by contraction,

(1.47) $$R_{jk} = R^s{}_{jks}$$

moreover, from R_{jk}, by multiplication by g^{jk} and by contraction, we get

---

[1]Some writers denote our $-R^i{}_{jkl}$ by $R^i{}_{jkl}$.

$$R = g^{jk}R_{jk} \tag{1.48}$$

R_{jk} and R are called "Ricci tensor" and "curvature scalar" respectively.

From the definition (1.44) of $R^i{}_{jkl}$, it is easily seen that $R^i{}_{jkl}$ satisfies the following algebraic identities:

$$R^i{}_{jkl} = -R^i{}_{jlk} \tag{1.49}$$

$$R^i{}_{jkl} + R^i{}_{klj} + R^i{}_{ljk} = 0 \tag{1.50}$$

and consequently, if we put

$$R_{ijkl} = g_{is}R^s{}_{jkl} \tag{1.51}$$

then R_{ijkl} satisfies

$$R_{ijkl} = -R_{ijlk} \tag{1.52}$$

$$R_{ijkl} + R_{iklj} + R_{iljk} = 0 \tag{1.53}$$

Equations (1.50) and (1.53) are called the first Bianchi identities.

Moreover, applying the Ricci formula to g_{ij}, we get

$$0 = g_{ij;k;l} - g_{ij;l;k} = -g_{sj}R^s{}_{ikl} - g_{is}R^s{}_{jkl}$$

from which

$$R_{ijkl} = -R_{jikl} \tag{1.54}$$

Calculating the covariant components R_{ijkl} explicitly, we find

$$R_{ijkl} = \frac{1}{2}\left(\frac{\partial^2 g_{ik}}{\partial x^j \partial x^l} + \frac{\partial^2 g_{jl}}{\partial x^i \partial x^k} - \frac{\partial^2 g_{jk}}{\partial x^i \partial x^l} - \frac{\partial^2 g_{il}}{\partial x^j \partial x^k} \right) - g_{rs}\left(\left\{{r \atop jk}\right\}\left\{{s \atop il}\right\} - \left\{{r \atop jl}\right\}\left\{{s \atop ik}\right\}\right) \tag{1.55}$$

which shows that

$$R_{ijkl} = R_{klij} \tag{1.56}$$

From (1.50), on contracting with respect to i and l, we obtain

$$R_{jk} - R_{kj} = 0 \tag{1.57}$$

by virtue of (1.49) and (1.54), and equation (1.57) shows that the Ricci tensor R_{jk} is a symmetric tensor.

It is to be noted that

$$g^{jk}R^i{}_{jkl} = g^{jk}g^{is}R_{sjkl} = g^{jk}g^{is}R_{jslk} = g^{is}R_{sl}$$

or

$$g^{jk}R^i{}_{jkl} = g^{is}R_{sl} \ (= R^i{}_l) \tag{1.58}$$

For the covariant derivative $R^i{}_{jkl;m}$ of the curvature tensor $R^i{}_{jkl}$, we can prove

$$R^i{}_{jkl;m} + R^i{}_{jlm;k} + R^i{}_{jmk;l} = 0 \tag{1.59}$$

which is called the second Bianchi identity. From (1.59), by contraction with respect to i and m, we find

$$R^s{}_{jkl;s} = R_{jk;l} - R_{jl;k} \tag{1.60}$$

and on multiplying this by g^{jk} and contracting, we get

$$2R^s{}_{l;s} = R_{;l} \tag{1.61}$$

5. SECTIONAL CURVATURE

In a two-dimensional Riemannian manifold whose fundamental quadratic form is

$$ds^2 = g_{11}(dx^1)^2 + 2g_{12}dx^1dx^2 + g_{22}(dx^2)^2$$

the only non-zero components of the Riemann-Christoffel curvature tensor R_{ijkl} are

$$R_{1212} = - R_{1221} = - R_{2112} = R_{2121}$$

The Gaussian curvature K of this manifold is then defined by

$$K = - \frac{R_{1212}}{g} \tag{1.62}$$

Now, in an n-dimensional Riemannian manifold, consider two contravariant vectors λ^i and μ^i at a point (x^i). These two vectors span a two-dimensional plane passing through the point (x^i). We next consider all the geodesics which pass through this point and are tangent to the two-dimensional plane spanned by λ^i and μ^i. These geodesics describe a two-dimensional surface which passes through the point (x^i) and is tangent to the two-dimensional plane spanned by λ^i and μ^i.

By virtue of (1.62), the Gaussian curvature K of this surface at (x^i) is given by

$$K = - \frac{R_{ijkl}\lambda^i\mu^j\lambda^k\mu^l}{(g_{ik}g_{jl} - g_{il}g_{jk})\lambda^i\mu^j\lambda^k\mu^l}$$

or

$$K = \frac{R_{ijkl}\lambda^i\mu^j\lambda^k\mu^l}{(g_{jk}g_{il} - g_{jl}g_{ik})\lambda^i\mu^j\lambda^k\mu^l} \tag{1.63}$$

This is called the sectional curvature at (x^i) with respect to the two-dimensional plane spanned by λ^i and μ^i.

If we choose λ^i and μ^i as two mutually orthogonal unit vectors in the two-dimensional plane, then equation (1.63) becomes

$$K = - R_{ijkl}\lambda^i\mu^j\lambda^k\mu^l \tag{1.64}$$

Now, suppose that, at a fixed point of our Riemannian manifold, this sectional curvature does not depend on the two-dimensional section passing through this point. Then, as we can see from (1.63), the curvature tensor R_{ijkl} must have the form

$$R_{ijkl} = K(g_{jk}g_{il} - g_{jl}g_{ik}) \tag{1.65}$$

and $R^i{}_{jkl}$ the form

(1.66) $$R^{i}{}_{jkl} = K(g_{jk}\delta^{i}_{l} - g_{jl}\delta^{i}_{k})$$

If, at any point of the manifold, this sectional curvature does not depend on the two-dimensional section passing through this point, then equations (1.65) and (1.66) must be valid at every point of the manifold, K being thus a function of the point.

From (1.66), by contracting with respect to i and l, we get

(1.67) $$R_{jk} = (n - 1)Kg_{jk}$$

from which, by multiplying g^{jk} and contracting, we find

(1.68) $$R = n(n - 1)K$$

Now, (1.67) can also be written as

$$R^{i}{}_{k} = \frac{1}{n} R \,\delta^{i}_{k}$$

and if we substitute this into (1.61), we find

$$\frac{2}{n} R_{;l} = R_{;l}$$

from which we can see that, for $n > 2$, R and consequently K are absolute constants.

Thus, if the sectional curvature at every point of the manifold does not depend on the two-dimensional planes passing through the point, then this sectional curvature is an absolute constant in the whole manifold. Such a Riemannian manifold is said to be of constant curvature.

If this constant is zero, then we have

(1.69) $$R^{i}{}_{jkl} = 0$$

In this case, the equations

$$\frac{\partial^2 x'^i}{\partial x^s \partial x^t} = \frac{\partial x'^i}{\partial x^r} \left\{ {r \atop st} \right\}$$

obtained from (1.22) by putting $\left\{ {i \atop jk} \right\}' = 0$ are completely integrable, and consequently there exists a coordinate system in which $\left\{ {i \atop jk} \right\}' = 0$, and consequently $g'_{jk} = \text{const}$. Thus, every coordinate neighborhood of the manifold can be mapped isometrically on a certain domain in the Euclidean space.

Conversely, if every coordinate neighborhood can be mapped isometrically on a certain domain in the Euclidean space, then it is evident that we have (1.69).

Such a Riemannian manifold is said to be locally Euclideen or to be locally flat.

Returning tó a general Riemannian manifold, we consider n contravariant mutually orthogonal unit vectors

$$\lambda_a^i \quad (a, b, c, \ldots = \dot{1}, \dot{2}, \ldots, \dot{n})$$

at a point (x^i). We then have

$$g_{ij}\lambda_a^i\lambda_b^j = \delta_{ab}$$

and therefore

$$g^{ij} = \sum_{a=1}^{n} \lambda_a^i \lambda_a^j \tag{1.70}$$

Now, the sectional curvature at this point, as determined by a two-dimensional plane spanned by λ_a^i and λ_b^i, is given by

$$K_{ab} = - R_{ijkl}\lambda_a^i\lambda_b^j\lambda_a^k\lambda_b^l$$

and hence

$$\sum_{b=1}^{n} K_{ab} = - R_{ijkl}\lambda_a^i\lambda_a^k g^{jl}$$

or

$$\sum_{b=1}^{n} K_{ab} = R_{jk}\lambda_a^j\lambda_a^k \tag{1.71}$$

and

$$\sum_{a=1}^{n}\sum_{b=1}^{n} K_{ab} = R \tag{1.72}$$

Formula (1.71) shows that, if we take a unit contravariant vector λ^i and consider $n - 1$ sectional curvatures determined by $n - 1$ two-dimensional planes spanned by λ^i and $n - 1$ unit vectors which are orthogonal to λ^i and to each other, then the sum of these $n - 1$ sectional curvatures is equal to $R_{jk}\lambda^j\lambda^k$ and is independent of the choice of other $n - 1$ orthogonal unit vectors. We call $R_{jk}\lambda^j\lambda^k$ the Ricci curvature with respect to the unit vector λ^i.

Equation (1.72) shows that the sum of n Ricci curvatures with respect to n mutually orthogonal unit vectors is equal to R and is independent of the choice of these n mutually orthogonal unit vectors.

Now, consider the Ricci curvature M with respect to a certain contravariant vector λ^i :

$$M = \frac{R_{jk}\lambda^j\lambda^k}{g_{jk}\lambda^j\lambda^k}$$

The direction which gives the extremum of M is given by

$$(1.73) \qquad (R_{jk} - Mg_{jk})\lambda^k = 0$$

and in general, there are n such directions which are mutually orthogonal. We call these directions Ricci directions.

A manifold for which the Ricci direction is indeterminate is called an Einstein manifold. For such a manifold, we have

$$(1.74) \qquad R_{jk} = Mg_{jk}$$

By multiplication by g^{jk} and contraction, we obtain

$$R = nM$$

from which

$$(1.75) \qquad R_{jk} = \frac{1}{n} Rg_{jk}$$

It will be easily seen from (1.61) that R is an absolute constant.

6. PARALLEL DISPLACEMENT

If v^i is a contravariant vector at a point (x^i) and $v^i + dv^i$ its value at an infinitesimally nearby point $(x^i + dx^i)$, then we know that

$$(1.76) \qquad \delta v^i = dv^i + \left\{{i \atop jk}\right\}v^j dx^k$$

are components of a contravariant vector.

If $\delta v^i = 0$, we say that the vector v^i at (x^i) and the vector $v^i + dv^i$ at $(x^i + dx^i)$ are parallel to each other, or that the vector $v^i + dv^i$ at $(x^i + dx^i)$ has been obtained from v^i at (x^i) by a parallel displacement. This definition is invariant relative to changes of coordinates. A similar definition applies to any tensor.

If we compare the equations

$$\frac{\delta v^i}{dt} = \frac{dv^i}{dt} + \left\{ {i \atop jk} \right\} v^j \frac{dx^k}{dt} \tag{1.77}$$

for the parallel displacement of the vector $v^i(t)$ along a curve $x^i(t)$ with the differential equations of geodesic

$$\frac{d^2x^i}{ds^2} + \left\{ {i \atop jk} \right\} \frac{dx^j}{ds} \frac{dx^k}{ds} = 0$$

then we see that the tangent dx^i/ds of a geodesic is displaced parallelly along the geodesic.

Since we have $\delta g_{jk} = 0$, it is easily seen that the length of a vector and the angle between two vectors are invariant by parallel displacements of these vectors.

If we want to displace parallelly a vector v_0^i at a point $P_0(x_0^i)$ to a point $P_1(x_1^i)$ which is at a finite distance from P_0, we must first assign a curve $x^i(t)$ joining two points P_0 and P_1 (and consequently, a curve $x^i(t)$ such that $x^i(t_0) = x_0^i$ and $x^i(t_1) = x_1^i$) and integrate the differential equations (1.77) with initial conditions $v^i(t_0) = v_0^i$. If we denote the solution by $v^i(t)$, then $v^i(t_1)$ is the vector which we get when we displace the vector v_0^i at the point $P_0(x_0^i)$ parallelly along the curve $x^i(t)$ to the point $P_1(x_1^i)$.

Thus, the parallelism depends on the curve which joins the starting point and the finishing point.

If the parallelism of a vector does not depend on the curve joining the starting point and the finishing point, then, at every point of the manifold, we have one and only one vector $v^i(x)$ which is parallel to the given vector v_0^i at the point $P_0(x_0^i)$, and the differential equations

$$\frac{\delta v^i}{dt} = v^i{}_{;k} \frac{dx^k}{dt} = 0$$

should be satisfied for any curve. Thus we have

$$v^i{}_{;k} = 0$$

from which, by virtue of (1.43),

$$v^j R^i{}_{jkl} = 0$$

Thus, if the parallelism of any vector does not depend on the curve along which the vector is displaced, then, the above equation having to be satisfied for any v^i , we must have

$$R^i{}_{jkl} = 0$$

and consequently, the manifold must be locally Euclidean.

CHAPTER II

HARMONIC AND KILLING VECTORS

1. THEOREM OF E. HOPF

In an n-dimensional coordinate neighborhood U, we consider a linear partial differential expression of the second order of elliptic type

$$L(\phi) = g^{jk} \frac{\partial^2 \phi}{\partial x^j \partial x^k} + h^i \frac{\partial \phi}{\partial x^i}$$

where $g^{jk}(x)$ and $h^i(x)$ are continuous functions of point $P(x)$ in U, and the quadratic form $g^{jk} z_j z_k$ is supposed to be positive definite everywhere in U.

We shall prove an important theorem due to E. Hopf [1]:

THEOREM 2.1. In a coordinate neighborhood U, if a function $\phi(P)$ of class C^2 satisfies the inequality $L(\phi) \geq 0$, and if there exists a fixed point P_0 in U such that $\phi(P) \leq \phi(P_0)$ everywhere in U, then we must have $\phi(P) = \phi(P_0)$ everywhere in U. If $L(\phi) \leq 0$ and $\phi(P) \geq \phi(P_0)$ everywhere in U, then we must have $\phi(P) = \phi(P_0)$ everywhere in U.

To prove the first part of the Theorem, we assume that

$$\phi(P) \not\equiv M$$

where $M = \phi(P_0)$, and we draw a contradiction from it.

Regarding (x^i) as coordinates of a point in an n-dimensional Euclidean domain U, we use hereafter the terminologies of Euclidean geometry.

As we have assumed that $\phi(P) \not\equiv M$ in U, there exists a point C in U such that $\phi(C) < M$, and if we describe a sphere with C as center and with a sufficiently small radius, then we have $\phi(P) < M$ in the entire sphere.

Next, increasing the radius of this sphere, we can get a sphere such that we have

$$\phi(P_1) = M$$

at a certain point P_1 which is on the intersection of U and the surface of the sphere and

$$\phi(P) < M$$

in the intersection of U and the interior of the sphere.

We now consider a sphere S which is tangent to the above-mentioned sphere at P_1 and which lies at the same time in U and in the above-mentioned sphere, and we shall denote the radius of this sphere by R.

We then have

$$\phi(P_1) = M \tag{2.1}$$

at only one point P_1 on the surface of S and

$$\phi(P) < M \tag{2.2}$$

at other points on the surface of S and inside of S.

We next consider a sphere S_1 whose center is P_1 and whose radius is

$$R_1 < R \tag{2.3}$$

and which lies moreover entirely inside of U. The surface of S_1 is divided into two parts by S. We denote the closure of the part which is inside of S by F_i and the closure of the part which is outside of S by F_0. On F_i we have $\phi < M$, and consequently $\phi \leq M - \epsilon$ for some fixed $\epsilon > 0$. Thus,

$$\begin{cases} \phi \leq M - \epsilon & \text{on } F_i \\ \phi \leq M & \text{on } F_0 \end{cases} \tag{2.4}$$

We now take the center of the sphere S as the origin of the orthogonal coordinate system and consider the function

$$\psi(P) = e^{-\alpha r^2} - e^{-\alpha R^2}$$

where α is a positive constant and

$$r^2 = (x^1)^2 + (x^2)^2 + \dots + (x^n)^2$$

and on applying the operator L to the function ψ, we find

$$L(\psi) = e^{-\alpha r^2}[4\alpha^2 g^{jk}x^j x^k - 2\alpha(h^i x^i + g^{ii})]$$

Since $R_1 < R$, the origin of the coordinate system, which is the center of the sphere S, is outside of the sphere S_1. Thus, on the surface of S_1 and inside of S_1, we have

$$g^{jk}x^j x^k > 0$$

and consequently

$$g^{jk}x^j x^k \geq \text{const.} > 0$$

Consequently, taking α large enough, we may assume that

$$(2.5) \qquad L(\psi) > 0 \qquad \text{in } S_1$$

On the other hand, we have

$$(2.6) \qquad \begin{cases} \psi(P) \leq 0 & \text{on } F_0 \\ \psi(P_1) = 0 \end{cases}$$

Finally, we put

$$\Phi(P) = \phi(P) + \delta\cdot\psi(P)$$

where δ is a positive small number chosen in such a way that

$$\Phi(P) < M \qquad \text{on } F_1$$

and this choice is possible by virtue of the first equation of (2.4).

By (2.4) and (2.6), we have

$$\Phi(P) < M \qquad \text{on } F_0$$

and therefore, on the whole boundary of S_1, we have

$$\Phi(P) < M$$

But, by (2.1) and (2.6), at the center P_1 of S_1, we have

$$\Phi(P_1) = M$$

Consequently, the function $\Phi(P)$ attains the maximum at a point which is inside of S_1. But this is impossible, since, in consequence of

$$L(\phi) \geq 0 \qquad \text{in } S_1$$

and

$$L(\psi) > 0 \qquad \text{in } S_1$$

we have

$$L(\Phi) > 0 \qquad \text{in } S_1 \tag{2.7}$$

Now at a point where the function Φ attains the maximum, $L(\Phi)$ reduces to

$$L(\Phi) = g^{jk} \frac{\partial^2 \Phi}{\partial x^j \partial x^k}$$

and we must have

$$\frac{\partial^2 \Phi}{\partial x^j \partial x^k} \lambda^j \lambda^k \leq 0$$

for any λ^i, and consequently, $g^{jk} z_j z_k$ being positive definite and

$$\frac{\partial^2 \Phi}{\partial x^j \partial x^k} \lambda^j \lambda^k$$

being negative definite, we must have

$$L(\Phi) = g^{jk} \frac{\partial^2 \Phi}{\partial x^j \partial x^k} \leq 0$$

which contradicts (2.7).

Thus the first part of the theorem is proved. The second part will be proved in a similar way.

Now, in a compact manifold V_n, suppose that a function $\phi(x)$ of class C^2 satisfies

$$L(\phi) = g^{jk} \frac{\partial^2 \phi}{\partial x^j \partial x^k} + h^i \frac{\partial \phi}{\partial x^i}$$

everywhere in V_n. Since the manifold is compact and the function $\phi(x)$ is continuous in this compact manifold, there exists a point P_0 at which the function ϕ attains the maximum, that is

$$\phi(P) \leq \phi(P_0) \tag{2.8}$$

everywhere in V_n. Therefore by Theorem 2.1 we have

$$\phi(P) \leq \phi(P_0) = M$$

in a certain neighborhood of P_0. But the points where $\phi(P)$ reaches its maximum form a closed set, and thus we attain the following conclusion.

THEOREM 2.2. In a compact space V_n, if a function $\phi(x)$ satisfies

$$L(\phi) = g^{jk}(x)\frac{\partial^2\phi}{\partial x^j \partial x^k} + h^i(x)\frac{\partial\phi}{\partial x^i} \geq 0$$

everywhere in V_n, where $g^{jk}(x)$ are coefficients of a positive definite quadratic form at any point of V_n, then we have

$$\phi = \text{const}$$

everywhere in V_n.

Moreover, since in a compact Riemannian manifold V_n with positive definite metric $ds^2 = g_{jk}dx^j dx^k$, we have

$$\Delta\phi = g^{jk}\phi_{;j;k} = g^{jk}\frac{\partial^2\phi}{\partial x^j \partial x^k} - g^{jk}\left\{ {i \atop jk} \right\}\frac{\partial\phi}{\partial x^i} \tag{2.9}$$

we can state the so-called Bochner's lemma:

THEOREM 2.3. In a compact Riemannian manifold with positive definite metric, if a function $\phi(x)$ satisfies

$$\Delta\phi \geq 0$$

everywhere in the manifold, then we have $\phi = \text{const}$

$$\Delta\phi = 0$$

everywhere in the manifold. (Bochner [2]).

2. THEOREM OF GREEN

The divergence of a contravariant vector λ^i being given by

$$\lambda^i{}_{;i} = \frac{1}{\sqrt{g}} \frac{\partial \sqrt{g}\, \lambda^i}{\partial x^i} \tag{2.10}$$

in the case of compact orientable Riemannian manifold, we can state the theorem of Green (see Bochner [1]):

THEOREM 2.4. In a compact orientable Riemannian manifold V_n, we have

$$\int_{V_n} \lambda^i{}_{;i} dv = 0 \tag{2.11}$$

for an arbitrary vector field $\lambda^i(x)$.

To prove this, we remark first that, if a bounded set D is contained in a coordinate neighborhood, then we have

$$\int_D \lambda^i{}_{;i} dv = \int_D \frac{\partial \sqrt{g}\lambda^i}{\partial x^i} dx^1 dx^2 \ldots dx^n$$

Suppose now that A is a "rectangle": $a^i < x^i < b^i$ and that λ^i vanishes on the boundary of A. In this case, we have

$$\int_{a^1}^{b^1} \frac{\partial\sqrt{g}\lambda^1}{\partial x^1} dx^1 = \int_{a^2}^{b^2} \frac{\partial\sqrt{g}\lambda^2}{\partial x^2} dx^2 = \ldots = \int_{a^n}^{b^n} \frac{\partial\sqrt{g}\lambda^n}{\partial x^n} dx^n = 0$$

and therefore

$$\int_A \lambda^i{}_{;i} dv = 0 \tag{2.12}$$

But, since the integral of $\lambda^i{}_{;i}$ is zero over any open set on which λ^i vanishes, equation (2.12) shows that (2.11) is true if λ^i

vanishes outside some "rectangle" A.

Now, since the manifold is compact, we can cover it by finite number of neighborhoods $U_1, U_2, \ldots, U_M$, whose closures are contained in "rectangles" $A_1, A_2, \ldots, A_M$ respectively. Corresponding to each α, $\alpha = 1, 2, \ldots, M$, we can easily find a neighborhood V_α between U_α and A_α and a non-negative scalar function ϕ_α of class C^1 in A_α such that $\phi_\alpha \geq 1$ in U_α and $\phi_\alpha = 0$ outside V_α. Completing the function ϕ_α by values zero outside A_α, we have, throughout V_n,

$$\phi_1 + \phi_2 + \cdots + \phi_M \geq 1$$

Thus, if we put

$$\psi_\alpha = \frac{\phi_\alpha}{\phi_1 + \phi_2 + \cdots + \phi_M}$$

then, the function ψ_α is of class C^1 and has the following property: ψ_α vanishes outside the "rectangle" A_α and

$$\psi_1 + \psi_2 + \cdots + \psi_M = 1$$

Hence, if we put

$$\lambda^i_\alpha = \psi_\alpha \lambda^i$$

then the contravariant vector field λ^i_α has the property that it vanishes outside the "rectangle" A_α. Thus we have

$$\int \lambda^i_{\alpha;i}\, dv = 0$$

But, on the other hand, we have

$$\lambda^i = \sum_{\alpha=1}^{M} \lambda^i_\alpha$$

and consequently

$$\lambda^i_{\ ;i} = \sum_{\alpha=1}^{M} \lambda^i_{\alpha;i}$$

Integrating this over the whole manifold, we have

$$\int \lambda^i_{\ ;i}\, dv = \sum_{\alpha=1}^{M} \int \lambda^i_{\alpha;i}\, dv = 0$$

which proves Theorem 2.4.

Since the Laplacean $\Delta\phi$ of a scalar field $\phi(x)$ can be written as

$$\Delta\phi = \phi^{i}{}_{;i} = (g^{ij}\phi_{;j})_{;i}$$

Theorem 2.4 implies as follows.

THEOREM 2.5. In a compact orientable Riemannian manifold V_n, for any scalar field $\phi(x)$, we have

$$\int_{V_n} \Delta\phi \, dv = 0 \tag{2.13}$$

If we apply the operator Δ to ϕ^2, then we get

$$\Delta\phi^2 = 2\phi\Delta\phi + 2g^{ij}\phi_{;i}\phi_{;j}$$

and consequently, on applying Theorem 2.5 to the scalar field ϕ^2, we obtain

$$\int_{V_n} (\phi\Delta\phi + g^{ij}\phi_{;i}\phi_{;j})dv = 0 \tag{2.14}$$

Now, if we have $\Delta\phi \geq 0$ everywhere in V_n, then, as is seen from (2.13), we must have $\Delta\phi = 0$ everywhere in V_n. Hence, as is seen from (2.14), we must have $g^{ij}\phi_{;i}\phi_{;j} = 0$ or $\phi_{;i} = 0$, or $\phi = \text{const.}$ This gives another proof of Theorem 2.3 in case the manifold is orientable.

3. SOME APPLICATIONS OF THE THEOREM OF HOPF-BOCHNER

In this section, we assume that the manifold is of class C^3 and g_{jk} of class C^2.

We consider a vector field $\xi^i(x)$ of class C^2 and we put

$$\phi = \xi^i\xi_i \qquad (\xi_i = g_{ij}\xi^j) \tag{2.15}$$

and for the Laplacean of the latter we have

$$\Delta\phi = 2(\xi^{i;j}\xi_{i;j} + \xi^i g^{bc}\xi_{i;b;c})$$

where we have put

$$\xi^{i;j} = \xi^{i}{}_{;a} g^{aj}$$

Now

$$\xi^{i;j}\xi_{i;j} = g^{ab}g^{cd}\xi_{a;c}\xi_{b;d}$$

is a positive definite form in $\xi_{a;c}$, and therefore if ξ_i satisfies equations of the form

$$g^{bc}\xi_{i;b;c} = T_{ij}\xi^j \tag{2.16}$$

and if the quadratic form $T_{ij}\xi^i\xi^j$ satisfies

$$T_{ij}\xi^i\xi^j \geq 0$$

then we have

$$\Delta\Phi = 2(\xi^{i;j}\xi_{i;j} + T_{ij}\xi^i\xi^j) \geq 0$$

Consequently, from Theorem 2.3, we get

$$\xi^{i;j}\xi_{i;j} + T_{ij}\xi^i\xi^j = 0$$

or

$$\xi_{i;j} = 0$$

and also $T_{ij}\xi^i\xi^j = 0$, and if the quadratic form $T_{ij}\xi^i\xi^j$ is positive definite, then we can conclude from $T_{ij}\xi^i\xi^j = 0$ that

$$\xi^i = 0$$

Thus we have

THEOREM 2.6. In a compact Riemannian manifold V_n, there exists no vector field which satisfies relations

$$g^{bc}\xi_{i;b;c} = T_{ij}\xi^j$$

and

$$T_{ij}\xi^i\xi^j \geq 0$$

unless we have

$$\xi_{i;j} = 0$$

and then automatically $T_{ij}\xi^i\xi^j = 0$.

Thus, the only exceptions are parallel vector fields, and there are no such vectors other than zero vectors if the quadratic form $T_{ij}\xi^i\xi^j$ is positive definite. (Bochner [10]).

We now take an arbitrary vector field $\xi_b(x)$ of class C^2 in V_n and write down the Ricci identity:

$$\xi_{b;i;c} - \xi_{b;c;i} = -\xi_a R^a{}_{bic}$$

from which we obtain

$$\xi_{i;b;c} - (\xi_{i;b} - \xi_{b;i})_{;c} - \xi_{b;c;i} = -\xi_a R^a{}_{bic}$$

or, multiplying by g^{bc} and contracting,

$$g^{bc}\xi_{i;b;c} - g^{bc}(\xi_{i;b} - \xi_{b;i})_{;c} - \xi^a{}_{;a;i} = R_{ai}\xi^a$$

Thus, if the vector field ξ_i satisfies

(2.17) $$g^{bc}(\xi_{i;b} - \xi_{b;i})_{;c} + \xi^a{}_{;a;i} = 0$$

then it satisfies also

(2.18) $$g^{bc}\xi_{i;b;c} = R_{ai}\xi^a$$

and, consequently, by Theorem 2.6, we have

THEOREM 2.7. In a compact Riemannian manifold V_n there exists no vector field which satisfies (2.17) and

$$R_{ij}\xi^i\xi^j \geq 0$$

unless we have

$$\xi_{i;j} = 0$$

and then automatically $R_{ij}\xi^i\xi^j = 0$.

Especially, if the manifold has positive definite Ricci curvature throughout, there exists no vector field other than zero vector which satisfies (2.17).

Next, we take again an arbitrary vector field $\xi_b(x)$ of class C^2 and write down the Ricci identity:

$$\xi_{b;i;c} - \xi_{b;c;i} = - \xi_a R^a{}_{bic}$$

from which we obtain

$$- \xi_{i;b;c} + (\xi_{i;b} + \xi_{b;i})_{;c} - \xi_{b;c;i} = - \xi_a R^a{}_{bic}$$

or, multiplying by g^{bc} and contracting,

$$- g^{bc}\xi_{i;b;c} + g^{bc}(\xi_{i;b} + \xi_{b;i})_{;c} - \xi^a{}_{;a;i} = + R_{ai}\xi^a$$

Thus, if the vector field ξ_i satisfies

(2.19) $$g^{bc}(\xi_{i;b} + \xi_{b;i})_{;c} - \xi^a{}_{;a;i} = 0$$

then it satisfies also

(2.20) $$g^{bc}\xi_{i;b;c} = - R_{ai}\xi^a$$

and, consequently, by Theorem 2.6, we have

THEOREM 2.8. In a compact Riemannian manifold V_n , there exists no vector field which satisfies (2.19) and

$$R_{ij}\xi^i\xi^j \leq 0$$

unless we have

$$\xi_{i;j} = 0$$

and then automatically $R_{ij}\xi^i\xi^j = 0$.

Especially, if the manifold has negative definite Ricci curvature throughout, there exists no vector field other than zero vector which satisfies (2.19).

4. HARMONIC VECTORS

A vector ξ_i is called harmonic if it satisfies the conditions

(2.21) $$\xi_{i;j} - \xi_{j;i} = 0$$

and

(2.22) $$\xi^i{}_{;i} = 0$$

It is well known that, in a compact orientable Riemannian manifold, the number of linearly independent (with constant coefficients) harmonic vectors is equal to the one-dimensional Betti number B_1 of the manifold, (Hodge [1]).

Now, if ξ_i is a harmonic vector, then it satisfies (2.17) and consequently we have (2.18). Thus, as a special case of Theorem 2.7, we can state

THEOREM 2.9. In a compact Riemannian manifold V_n , there exists no harmonic vector field which satisfies

$$R_{ij}\xi^i\xi^j \geq 0$$

unless we have

$$\xi_{i;j} = 0$$

and then automatically $R_{ij}\xi^i\xi^j = 0$.

Especially, if the manifold has positive definite Ricci curvature throughout, there exists no harmonic vector other than zero vector and consequently, if the manifold is orientable, $B_1 = 0$. (Bochner [2], Myers [1]).

5. KILLING VECTORS

An infinitesimal point transformation

(2.23) $$\bar{x}^i = x^i + \xi^i(x)\delta t$$

is said to define an infinitesimal motion in V_n if the infinitesimal distance ds between two arbitrary points (x^i) and $(x^i + dx^i)$ is equal to the infinitesimal distance $d\bar{s}$ between two corresponding points $(\bar{x}^i)$ and $(\bar{x}^i + d\bar{x}^i)$, except for higher terms in δt.

Now, we have

$$ds^2 = g_{jk}(x)dx^j dx^k$$

and

$$d\bar{s}^2 = g_{jk}(\bar{x})d\bar{x}^j d\bar{x}^k$$

Thus, a necessary and sufficient condition that (2.23) be an infinitesimal motion of the manifold is that

$$g_{jk}(\bar{x})d\bar{x}^j d\bar{x}^k = g_{jk}(x)dx^j dx^k$$

or that

$$\left(g_{jk} + \xi^a \frac{\partial g_{jk}}{\partial x^a}\delta t\right)\left(dx^j + \frac{\partial \xi^j}{\partial x^b}dx^b \delta t\right)\left(dx^k + \frac{\partial \xi^k}{\partial x^c}dx^c \delta t\right) = g_{jk}(x)dx^j dx^k$$

be satisfied for any dx^i, except for higher terms in δt, that is to say, that

(2.24) $$\xi^a \frac{\partial g_{jk}}{\partial x^a} + \frac{\partial \xi^a}{\partial x^j} g_{ak} + \frac{\partial \xi^a}{\partial x^k} g_{ja} = 0$$

This equation is in tensor form:

$$\xi^a g_{jk;a} + \xi^a{}_{;j} g_{ak} + \xi^a{}_{;k} g_{ja} = 0$$

or

$$\xi_{j;k} + \xi_{k;j} = 0$$

and is called Killing's equation. We shall call a vector satisfying Killing's equation a Killing vector.

Now, if the manifold admits an infinitesimal motion (2.23), then the vector ξ_i satisfies (2.24). If we choose a coordinate system in which the vector ξ^i has the components

$$\xi^i = \delta^i_1$$

then, equation (2.24) becomes

$$\frac{\partial g_{jk}}{\partial x^1} = 0$$

which shows that the components g_{jk} of the fundamental tensor do not contain the variable x^1 in this special coordinate system. Thus the manifold admits a one-parameter group of motions

$$\bar{x}^i = x^i + \delta^i_1 . t$$

which is generated by ξ^i.

Now, if ξ^i is a Killing vector, then we have

$$\xi_{i;j} + \xi_{j;i} = 0$$

and automatically

$$\xi^i_{\ ;i} = 0$$

Thus, it satisfies (2.19), and consequently we have (2.20). Thus, as a special case of Theorem 2.8, we have

THEOREM 2.10. In a compact Riemannian manifold V_n there exists no Killing vector field which satisfies

$$R_{ij}\xi^i\xi^j \leq 0$$

unless we have

$$\xi_{i;j} = 0$$

and then automatically $R_{ij}\xi^i\xi^j = 0$.

Especially, if the manifold has negative definite Ricci curvature throughout, there exists no Killing vector field other than zero vector, and consequently there exists no one-parameter group of motions. (Bochner [2]).

6. AFFINE COLLINEATIONS

The geodesics in V_n are given by the differential equations

$$\frac{d^2x^i}{ds^2} + \Gamma^i_{jk}(x)\frac{dx^j}{ds}\frac{dx^k}{ds} = 0 \tag{2.26}$$

where $\Gamma^i_{jk} = \{{}^{\,i}_{jk}\}$ and s is the arc length.

An infinitesimal point transformation

$$\bar{x}^i = x^i + \xi^i(x)\delta t \tag{2.27}$$

is said to define an infinitesimal affine collineation in V_n, if the transformation (2.27) carries, infinitesimally, every geodesic of the manifold into a geodesic and if the arc length s receives an affine transformation.

Now, if the transformation (2.27) is an infinitesimal affine collineation, then it will carry the geodesic (2.26) into the geodesic

$$\frac{d^2\bar{x}^i}{d\bar{s}^2} + \Gamma^i_{jk}(\bar{x})\frac{d\bar{x}^j}{d\bar{s}}\frac{d\bar{x}^k}{d\bar{s}} = 0 \tag{2.28}$$

where

$$\bar{s} = as + b \tag{2.29}$$

a and b being constants.

From (2.28), we have

$$\frac{\partial^2\xi^i}{\partial x^j\partial x^k}\frac{dx^j}{ds}\frac{dx^k}{ds}\delta t + \left(\delta^i_a + \frac{\partial\xi^i}{\partial x^a}\delta t\right)\frac{d^2x^a}{ds^2}$$
$$+ \left(\Gamma^i_{jk} + \xi^l\frac{\partial\Gamma^i_{jk}}{\partial x^l}\delta t\right)\left(\delta^j_b + \frac{\partial\xi^j}{\partial x^b}\delta t\right)\left(\delta^k_c + \frac{\partial\xi^k}{\partial x^c}\delta t\right)\frac{dx^b}{ds}\frac{dx^c}{ds} = 0$$

from which, substituting (2.26), we obtain

$$\left(\frac{\partial^2\xi^i}{\partial x^j\partial x^k} + \xi^l\frac{\partial\Gamma^i_{jk}}{\partial x^l} - \frac{\partial\xi^i}{\partial x^a}\Gamma^a_{jk} + \frac{\partial\xi^a}{\partial x^j}\Gamma^i_{ak} + \frac{\partial\xi^a}{\partial x^k}\Gamma^i_{ja}\right)\frac{dx^j}{ds}\frac{dx^k}{ds} = 0$$

But, since the transformation (2.27) carries every geodesic into a geodesic, we must have

$$(2.30)\qquad \frac{\partial^2 \xi^i}{\partial x^j \partial x^k} + \xi^l \frac{\partial \Gamma^i_{jk}}{\partial x^l} - \frac{\partial \xi^i}{\partial x^a}\Gamma^a_{jk} + \frac{\partial \xi^a}{\partial x^j}\Gamma^i_{ak} + \frac{\partial \xi^a}{\partial x^k}\Gamma^i_{ja} = 0$$

or, in tensor form,

$$(2.31)\qquad \xi^i{}_{;j;k} + R^i{}_{jkl}\xi^l = 0$$

Also, if the manifold admits an infinitesimal affine collineation (2.27) then the vector ξ^i satisfies (2.30). If we choose a coordinate system in which the vector ξ^i has the components $\xi^i = \delta^i_1$, then equation (2.30) becomes

$$\frac{\partial \Gamma^i_{jk}}{\partial x^1} = 0$$

which shows that the Christoffel symbols $\Gamma^i_{jk} = \{^{\ i}_{jk}\}$ do not depend on the variable x^1 in this special coordinate system. Thus the manifold admits a one-parameter group of affine collineations

$$\bar{x}^i = x^i + \delta^i_1 . t$$

which is generated by (2.27).

Take a vector field ξ^i which satisfies (2.31). If we multiply by g^{jk} and contract, we obtain

$$g^{bc}\xi_{i;b;c} = - R_{ai}\xi^a$$

and thus from Theorem 2.6 we obtain

THEOREM 2.11. In a compact Riemannian manifold V_n , there exists no one-parameter group of affine collineations whose generating vector satisfies

$$R_{ij}\xi^i\xi^j \leq 0$$

unless we have

$$\xi_{i;j} = 0$$

and then automatically $R_{ij}\xi^i\xi^j = 0$.

Especially, if the manifold has negative definite

Ricci curvature throughout, there exists no one-parameter group of affine collineations in the manifold.

7. A THEOREM ON HARMONIC AND KILLING VECTORS

We know that, if ξ_i is a harmonic vector, then it satisfies

$$\xi_{i;j} = \xi_{j;i} \qquad \xi^a{}_{;a} = 0$$

and

$$g^{bc}\xi_{i;b;c} = R_{ij}\xi^j$$

and if η^i is a Killing vector, then it satisfies

$$\eta_{i;j} = - \eta_{j;i}$$

and

$$g^{bc}\eta^i{}_{;b;c} = - R^i{}_j\eta^j$$

If we apply the operator Δ to the inner product of these two vectors, we obtain

$$\Delta(\xi_i\eta^i) = g^{bc}\xi_{i;b;c}\eta^i + 2\xi_{i;j}\eta^{i;j} + \xi_i g^{bc}\eta^i{}_{;b;c}$$

but, on the other hand, we have

$$g^{bc}\xi_{i;b;c}\eta^i = R_{ij}\xi^i\eta^j$$

$$\xi_{i;j}\eta^{i;j} = 0$$

$$\xi_i g^{bc}\eta^i{}_{;b;c} = - R_{ij}\xi^i\eta^j$$

and consequently

$$\Delta(\xi_i\eta^i) = 0$$

Therefore, by Theorem 2.3,

$$\xi_i \eta^i = \text{constant} \tag{2.32}$$

and consequently

THEOREM 2.12. In a compact Riemannian manifold V_n, the inner product of a harmonic vector and a Killing vector is constant. (Bochner [8]).

8. LIE DERIVATIVES

We know that a necessary and sufficient condition that an infinitesimal point transformation

$$\bar{x}^i = x^i + \xi^i(x)\delta t \tag{2.33}$$

be an infinitesimal motion is that

$$g_{jk}(\bar{x})d\bar{x}^j d\bar{x}^k = g_{jk}(x)dx^j dx^k \tag{2.34}$$

be satisfied for any dx^i, except for higher terms in δt.

But, if we regard (2.33) as a coordinate transformation, then, $g_{jk}(x)dx^j dx^k$ being a scalar, we have

$$g_{jk}(x)dx^j dx^k = \bar{g}_{jk}(\bar{x})d\bar{x}^j d\bar{x}^k \tag{2.35}$$

where $\bar{g}_{jk}(\bar{x})$ are components of the fundamental metric tensor in the coordinate system $(\bar{x}^i)$, and consequently are given by

$$\bar{g}_{jk}(\bar{x}) = \frac{\partial x^b}{\partial \bar{x}^j} \frac{\partial x^c}{\partial \bar{x}^k} g_{bc}(x)$$

From (2.34) and (2.35), we have

$$g_{jk}(\bar{x}) - \bar{g}_{jk}(\bar{x}) = 0 \tag{2.36}$$

and thus on putting

$$Dg_{jk} = (Lg_{jk})\delta t = g_{jk}(\bar{x}) - \bar{g}_{jk}(\bar{x})$$

we have

$$Lg_{jk} = \xi^a \frac{\partial g_{jk}}{\partial x^a} + \frac{\partial \xi^a}{\partial x^j} g_{ak} + \frac{\partial \xi^a}{\partial x^k} g_{ja}$$

or

$$\mathcal{L}g_{jk} = \xi_{j;k} + \xi_{k;j} \tag{2.38}$$

We call $\mathcal{L}g_{jk}$ the Lie derivative of the tensor g_{jk} with respect to the infinitesimal point transformation (2.33), or with respect to the vector field ξ^i.

A necessary and sufficient condition that an infinitesimal point transformation (2.33) be a motion of the manifold is that the Lie derivative of the fundamental metric tensor with respect to (2.33) shall be zero.

On the other hand, in order to find a necessary and sufficient condition for (2.33) to be an affine collineation, we can proceed as follows:

The transformation (2.33) carries every geodesic

$$\frac{d^2x^i}{ds^2} + \Gamma^i_{jk}(x)\frac{dx^j}{ds}\frac{dx^k}{ds} = 0 \tag{2.39}$$

into the geodesic

$$\frac{d^2\bar{x}^i}{d\bar{s}^2} + \Gamma^i_{jk}(\bar{x})\frac{d\bar{x}^j}{d\bar{s}}\frac{d\bar{x}^k}{d\bar{s}} = 0$$

or

$$\frac{d^2\bar{x}^i}{ds^2} + \Gamma^i_{jk}(\bar{x})\frac{d\bar{x}^j}{ds}\frac{d\bar{x}^k}{ds} = 0 \tag{2.40}$$

Since the left hand side of (2.39) are components of a vector, if we regard (2.33) as a coordinate transformation, then equation (2.39) may be written as

$$\frac{d^2\bar{x}^i}{ds^2} + \bar{\Gamma}^i_{jk}(\bar{x})\frac{d\bar{x}^j}{ds}\frac{d\bar{x}^k}{ds} = 0 \tag{2.41}$$

in the coordinate system $(\bar{x}^i)$, where $\bar{\Gamma}^i_{jk}(\bar{x})$ are Christoffel symbols in the coordinate system $(\bar{x}^i)$ and consequently are given by

$$\bar{\Gamma}^i_{jk}(\bar{x}) = \frac{\partial \bar{x}^i}{\partial x^a}\left(\frac{\partial x^b}{\partial \bar{x}^j}\frac{\partial x^c}{\partial \bar{x}^k}\Gamma^a_{bc}(x) + \frac{\partial^2 x^a}{\partial \bar{x}^j \partial \bar{x}^k}\right) \tag{2.42}$$

Now, comparing (2.40) and (2.41), we get relations

$$\left(\Gamma^i_{jk}(\bar{x}) - \bar{\Gamma}^i_{jk}(\bar{x})\right)\frac{d\bar{x}^j}{ds}\frac{d\bar{x}^k}{ds} = 0$$

which must be satisfied by any $d\bar{x}^i/ds$, from which

$$\Gamma^i_{jk}\,(\bar{x}) - \bar{\Gamma}^i_{jk}\,(\bar{x}) = 0 \tag{2.43}$$

and for

$$D\Gamma^i_{jk} = (L\Gamma^i_{jk})\delta t = \Gamma^i_{jk}\,(\bar{x}) - \bar{\Gamma}^i_{jk}\,(\bar{x})$$

we obtain

$$L\Gamma^i_{jk} = \frac{\partial^2 \xi^i}{\partial x^j \partial x^k} + \xi^l \frac{\partial \Gamma^i_{jk}}{\partial x^l} - \frac{\partial \xi^i}{\partial x^a}\Gamma^a_{jk} + \frac{\partial \xi^a}{\partial x^j}\Gamma^i_{ak} + \frac{\partial \xi^a}{\partial x^k}\Gamma^i_{ja} \tag{2.44}$$

or

$$L\Gamma^i_{jk} = \xi^i{}_{;j;k} + R^i{}_{jkl}\xi^l \tag{2.45}$$

We call $L\Gamma^i_{jk}$ the Lie derivative of the affine connection Γ^i_{jk} with respect to the infinitesimal point transformation (2.33), or with respect to the vector field ξ^i .

We can see that a necessary and sufficient condition that an infinitesimal point transformation (2.33) be an infinitesimal affine collineation of the manifold is that the Lie derivative of the Christoffel symbols with respect to (2.33) vanish.

In general, when a field $\Omega(x)$ of a geometric object is given, we define the Lie derivative $L\Omega$ of Ω with respect to ξ^i by the equation

$$D\Omega = (L\Omega)\delta t = \Omega(\bar{x}) - \bar{\Omega}(\bar{x}) \tag{2.46}$$

where $\bar{\Omega}(\bar{x})$ denotes the components of this object in the coordinate system $(\bar{x}^i)$, (2.33) being regarded as a coordinate transformation from (x^i) to $(\bar{x}^i)$.

By a straightforward calculation, we can prove the following formulas:

For a contravariant vector v^i :

$$Lv^i = \xi^a v^i{}_{;a} - \xi^i{}_{;a} v^a \tag{2.47}$$

for a covariant vector v_j :

$$Lv_j = \xi^a v_{j;a} + \xi^a{}_{;j} v_a \tag{2.48}$$

for a mixed tensor, say, $T^i{}_{jk}$:

(2.49) $$LT^i{}_{jk} = \xi^a T^i{}_{jk;a} - \xi^i{}_{;a} T^a{}_{jk} + \xi^a{}_{;j} T^i{}_{ak} + \xi^a{}_{;k} T^i{}_{ja}$$

Now, for the fundamental metric tensor g_{aj} , we have

$$Lg_{aj} = \xi^b g_{aj;b} + \xi^b{}_{;a} g_{bj} + \xi^b{}_{;j} g_{ab}$$

and hence

$$Lg_{aj} = \xi_{a;j} + \xi_{j;a}$$

and this gives

$$(Lg_{aj})_{;k} = \xi_{a;j;k} + \xi_{j;a;k}$$

$$(Lg_{ak})_{;j} = \xi_{a;k;j} + \xi_{k;a;j}$$

$$-(Lg_{jk})_{;a} = -\xi_{j;k;a} - \xi_{k;j;a}$$

Adding these three, we find

$$(Lg_{aj})_{;k} + (Lg_{ak})_{;j} - (Lg_{jk})_{;a} = 2\xi_{a;j;k} + \xi_b R^b{}_{ajk} + \xi_b R^b{}_{jka} + \xi_b R^b{}_{kja}$$

$$= 2(\xi_{a;j;k} + R_{ajkl}\xi^l)$$

by virtue of

$$R^b{}_{ajk} + R^b{}_{jka} + R^b{}_{kaj} = 0$$

and

$$R_{bkja} = R_{ajkb}$$

Thus we have

$$\frac{1}{2} g^{ia}[(Lg_{aj})_{;k} + (Lg_{ak})_{;j} - (Lg_{jk})_{;a}] = \xi^i{}_{;j;k} + R^i{}_{jkl}\xi^l$$

or

(2.50) $$L\Gamma^i_{jk} = \frac{1}{2} g^{ia}[(Lg_{aj})_{;k} + (Lg_{ak})_{;j} - (Lg_{jk})_{;a}]$$

which shows that a motion in a Riemannian manifold is necessarily an affine collineation.

Next, for a contravariant vector field $v^i(x)$, we obtain, by a straightforward calculation,

$$L(v^i{}_{;k}) - (Lv^i)_{;k} = v^j(L\Gamma^i_{jk}) \tag{2.51}$$

Similarly, for a covariant vector field $v_j(x)$, we get

$$L(v_{j;k}) - (Lv_j)_{;k} = - v_i(L\Gamma^i_{jk}) \tag{2.52}$$

Finally, for a general tensor, say, $T^i{}_{jk}$, we obtain

$$L(T^i{}_{jk;l}) - (LT^i{}_{jk})_{;l} = T^a{}_{jk}(L\Gamma^i_{al}) - T^i{}_{ak}(L\Gamma^a_{jl}) - T^i{}_{ja}(L\Gamma^a_{kl}) \tag{2.53}$$

These equations show that a necessary and sufficient condition that covariant differentiation and Lie derivation be commutative is that the vector field ξ^i define an affine collineation.

Now, from

$$L\Gamma^i_{jk} = \xi^i{}_{;j;k} + R^i{}_{jkl}\xi^l$$

we find

$$(L\Gamma^i_{jk})_{;l} = \xi^i{}_{;j;k;l} + R^i{}_{jkm;l}\xi^m + R^i{}_{jkm}\xi^m{}_{;l}$$

and consequently

$$(L\Gamma^i_{jk})_{;l} - (L\Gamma^i_{jl})_{;k} = LR^i{}_{jkl} \tag{2.54}$$

and thus, for a motion, we have

$$LR^i{}_{jkl} = 0 \tag{2.55}$$

9. LIE DERIVATIVES OF HARMONIC TENSORS

A tensor $\xi_{i_1 i_2 \ldots i_p}$ is called harmonic, if it satisfies the conditions:

(2.56) $\xi_{i_1 i_2 \ldots i_p}$ is anti-symmetric in all the indices,

$$\xi_{[i_1 i_2 \cdots i_p ; j]} = 0 \tag{2.57}$$

or explicitly

$$\xi_{i_1 i_2 \cdots i_p ; j} = \xi_{j i_2 \cdots i_p ; i_1} + \xi_{i_1 j i_3 \cdots i_p ; i_2} + \cdots + \xi_{i_1 i_2 \cdots i_{p-1} j ; i_p} \tag{2.58}$$

and furthermore

$$g^{ij} \, \xi_{i i_2 \cdots i_p ; j} = 0 \tag{2.59}$$

It is well known that in a compact orientable Riemannian manifold, the number of linearly independent (with constant coefficients) harmonic tensors of order p is equal to the p-dimensional Betti number B_p of the manifold, (Hodge [1]).

Assume now that the manifold V_n admits a one parameter group of motions generated by

$$\bar{x}^i = x^i + \eta^i(x)\delta t$$

and put

$$Lf = \eta^i \frac{\partial}{\partial x^i} f$$

so that

$$Lg_{jk} = 0$$

and covariant differentiation and Lie derivation are commutative.

If we now apply the operator L to a harmonic tensor

$$\xi_{i_1 i_2 \cdots i_p}$$

then

(2.60) $L\xi_{i_1 i_2 \cdots i_p}$ is anti-symmetric in all the indices,

$$(L\xi_{i_1 i_2 \cdots i_p})_{;j} = (L\xi_{j i_2 \cdots i_p})_{;i_1} + (L\xi_{i_1 j i_3 \cdots i_p})_{;i_2} + \cdots + (L\xi_{i_1 i_2 \cdots i_{p-1} j})_{;i_p} \tag{2.61}$$

$$(2.62) \qquad g^{ij}(L\,\xi_{ii_2\ldots i_p})_{;j} = 0$$

and thus the Lie derivative

$$L\,\xi_{i_1 i_2 \ldots i_p}$$

is again a harmonic tensor.

But, on the other hand, we have by our general definition

$$L\,\xi_{i_1 i_2\ldots i_p} = \eta^a \xi_{i_1 i_2 \ldots i_p;a} + \eta^a{}_{;i_1}\xi_{ai_2\ldots i_p} + \ldots + \eta^a{}_{;i_p}\xi_{i_1 i_2 \ldots i_{p-1}a}$$

$$= \eta^a(\xi_{ai_2\ldots i_p;i_1} + \xi_{i_1 a i_3 \ldots i_p;i_2} + \ldots + \xi_{i_1 i_2 \ldots i_{p-1}a;i_p})$$

$$+ \eta^a{}_{;i_1}\xi_{ai_2\ldots i_p} + \eta^a{}_{;i_2}\xi_{i_1 a i_3\ldots i_p} + \ldots + \eta^a{}_{;i_p}\xi_{i_1 i_2\ldots i_{p-1}a}$$

$$= (\eta^a \xi_{ai_2\ldots i_p})_{;i_1} + (\eta^a \xi_{i_1 a i_3 \ldots i_p})_{;i_2} + \ldots + (\eta^a \xi_{i_1 i_2 \ldots i_{p-1}a})_{;i_p}$$

which shows that the harmonic differential form

$$(L\,\xi_{i_1 i_2 \ldots i_p})dx^{i_1} \wedge dx^{i_2} \wedge \ldots \wedge dx^{i_p}$$

is the exterior derivative of the form

$$p(\eta^a \xi_{ai_2\ldots i_p})dx^{i_2} \wedge \ldots \wedge dx^{i_p}$$

and since the harmonic form which is the exterior derivative of another form is identically zero, we obtain

THEOREM 2.13. If a compact orientable Riemannian manifold admits a one-parameter group of motions, then the Lie derivative of a harmonic tensor with respect to this group is identically zero. (Yano [3]).

Now, if there exist, in the manifold, a harmonic vector ξ_i and a Killing vector η^i, then, applying Theorem 2.13, we have

$$\begin{aligned} L\xi_i &= \eta^a \xi_{i;a} + \eta^a{}_{;i}\xi_a \\ &= \eta^a \xi_{a;i} + \eta^a{}_{;i}\xi_a \\ &= (\xi_a \eta^a)_{;i} \\ &= 0 \end{aligned}$$

from which we conclude

$$\xi_a \eta^a = \text{constant} \tag{2.63}$$

which gives another proof of Theorem 2.12 for an orientable manifold.

10. A FUNDAMENTAL FORMULA

In a compact orientable Riemannian manifold V_n we consider an arbitrary vector field $\xi^i(x)$ and we form the new vector field

$$\xi^i{}_{;j}\xi^j$$

whose divergence is

$$(\xi^i{}_{;j}\xi^j)_{;i} = \xi^i{}_{;j;i}\xi^j + \xi^i{}_{;j}\xi^j{}_{;i} \tag{2.64}$$

On the other hand, from the Ricci identity:

$$\xi^i{}_{;j;k} - \xi^i{}_{;k;j} = R^i{}_{ajk}\xi^a$$

we have, by contracting with respect to i and k,

$$\xi^i{}_{;j;i} - \xi^i{}_{;i;j} = R_{aj}\xi^a$$

or

$$\xi^i{}_{;j;i} = \xi^i{}_{;i;j} + R_{ij}\xi^i$$

and on substituting this into (2.64), we obtain

$$(\xi^i{}_{;j}\xi^j)_{;i} = \xi^i{}_{;i;j}\xi^j + R_{ij}\xi^i\xi^j + \xi^i{}_{;j}\xi^j{}_{;i} \tag{2.65}$$

Next we form the vector field

$$\xi^i{}_{;i}\xi^j$$

whose divergence is

$$(\xi^i{}_{;i}\xi^j)_{;j} = \xi^i{}_{;i;j}\xi^j + \xi^i{}_{;i}\xi^j{}_{;j} \tag{2.66}$$

and from (2.65) - (2.66), we obtain

$$(\xi^i{}_{;j}\xi^j)_{;i} - (\xi^i{}_{;i}\xi^j)_{;j} = R_{ij}\xi^i\xi^j + \xi^i{}_{;j}\xi^j{}_{;i} - \xi^i{}_{;i}\xi^j{}_{;j} \tag{2.67}$$

Integrating both members of (2.67) over the whole manifold, and applying Theorem 2.4, we obtain the formula

$$\int_{V_n}(R_{ij}\xi^i\xi^j + \xi^i{}_{;j}\xi^j{}_{;i} - \xi^i{}_{;i}\xi^j{}_{;j})dv = 0 \tag{2.68}$$

for which, on putting

$$\xi^{i;j} = \xi^i{}_{;a}g^{aj}$$

we can also write

$$\int_{V_n}(R_{ij}\xi^i\xi^j + \xi^{i;j}\xi_{j;i} - \xi^i{}_{;i}\xi^j{}_{;j})dv = 0 \tag{2.69}$$

(Yano [3]), and this formula, which is valid for any vector field $\xi^i(x)$, will be used extensively in the following discussions.

11. SOME APPLICATIONS OF THE FUNDAMENTAL FORMULA

First, if $\xi_i(x)$ is a harmonic vector field, then

$$\xi_{i;j} = \xi_{j;i} \qquad \xi^i{}_{;i} = 0$$

and consequently, the fundamental formula (2.69) gives

$$\int_{V_n}(R_{ij}\xi^i\xi^j + \xi^{i;j}\xi_{i;j})dv = 0 \tag{2.70}$$

But, since

$$\xi^{i;j}\xi_{i;j} = g^{ac}g^{bd}\xi_{a;b}\xi_{c;d}$$

and our metric is positive definite, we have

$$\xi^{i;j}\xi_{i;j} \geq 0$$

equality occurring when and only when $\xi_{i;j} = 0$; and thus, if

$$R_{ij}\xi^{i}\xi^{j} \geq 0$$

then, from (2.70), we conclude

$$R_{ij}\xi^{i}\xi^{j} = 0 \qquad \text{and} \qquad \xi_{i;j} = 0$$

Moreover, if $R_{ij}\xi^{i}\xi^{j}$ is a positive definite form, then from (2.70), we conclude

$$\xi_{i} = 0$$

and this gives another proof of Theorem 2.9 for an orientable manifold. (Yano [3]).

Next, if $\xi_{i}(x)$ is a Killing vector field, then

$$\xi_{i;j} + \xi_{j;i} = 0 \qquad \text{and automatically} \qquad \xi^{i}{}_{;i} = 0$$

and consequently, the fundamental formula (2.69) gives

$$(2.71) \qquad \int_{V_n}(R_{ij}\xi^{i}\xi^{j} - \xi^{i;j}\xi_{i;j})dv = 0$$

so that

$$R_{ij}\xi^{i}\xi^{j} \leq 0$$

implies

$$R_{ij}\xi^{i}\xi^{j} = 0 \qquad \text{and} \qquad \xi_{i;j} = 0$$

Moreover, if $R_{ij}\xi^i\xi^j$ is a negative definite form, then, from (2.71), we conclude

$$\xi_i = 0$$

and this gives another proof of Theorem 2.10 for an orientable manifold. (Yano [3]).

12. CONFORMAL TRANSFORMATIONS

An infinitesimal point transformation

$$\bar{x}^i = x^i + \xi^i(x)\delta t$$

is said to define an infinitesimal conformal transformation in V_n if the angle θ between two directions dx^i and δx^i at (x^i) is equal to the angle $\bar{\theta}$ between corresponding directions $d\bar{x}^i$ and $\delta\bar{x}^i$ at $(\bar{x}^i)$, neglecting higher terms in δt.

Now,

$$\cos\theta = \frac{g_{jk}(x)\, dx^j\delta x^k}{\sqrt{g_{jk}(x)dx^j dx^k}\sqrt{g_{jk}(x)\delta x^j\delta x^k}}$$

and

$$\cos\bar{\theta} = \frac{g_{jk}(\bar{x})\, d\bar{x}^j\delta\bar{x}^k}{\sqrt{g_{jk}(\bar{x})d\bar{x}^j d\bar{x}^k}\sqrt{g_{jk}(\bar{x})\delta\bar{x}^j\delta\bar{x}^k}}$$

and since the angle θ is a scalar, the first of these formulas can be written also in the form

$$\cos\theta = \frac{\bar{g}_{jk}(\bar{x})\, d\bar{x}^j\delta\bar{x}^k}{\sqrt{\bar{g}_{jk}(\bar{x})d\bar{x}^j d\bar{x}^k}\sqrt{\bar{g}_{jk}(\bar{x})\delta\bar{x}^j\delta\bar{x}^k}}$$

where $\bar{g}_{jk}(\bar{x})$ are components of the fundamental metric tensor in the coordinate system $(\bar{x}^i)$, and $\bar{x}^i = x^i + \xi^i(x)\delta t$ is regarded as a coordinate transformation $(x^i) \longrightarrow (\bar{x}^i)$.

Thus, a necessary and sufficient condition for $\bar{x}^i = x^i + \xi^i(x)\delta t$ to be an infinitesimal conformal transformation is

$$g_{jk}(\bar{x}) = (1 + 2\phi\delta t)\bar{g}_{jk}(\bar{x})$$

or

$$Dg_{jk} = g_{jk}(\bar{x}) - \bar{g}_{jk}(\bar{x}) = 2\phi g_{jk}\delta t$$

or

(2.72) $$Lg_{jk} = \xi_{j;k} + \xi_{k;j} = 2\phi g_{jk}$$

and if we assume that the vector field $\xi^i(x)$ defines an infinitesimal conformal transformation, then we have

$$\xi_{j;k} + \xi_{k;j} = 2\phi g_{jk} \qquad \text{and} \qquad \xi^i{}_{;i} = n\phi$$

Thus, the fundamental formula (2.69) gives

$$\int_{V_n} [R_{ij}\xi^i\xi^j + \xi^{i;j}(2\phi g_{ij} - \xi_{i;j}) - n^2\phi^2]dv = 0$$

or

(2.73) $$\int_{V_n} [R_{ij}\xi^i\xi^j - \xi^{i;j}\xi_{i;j} - n(n - 2)\phi^2]dv = 0$$

and consequently, if

$$R_{ij}\xi^i\xi^j \leq 0$$

then we must have, for $n \geq 2$,

$$R_{ij}\xi^i\xi^j = 0 \qquad \xi_{i;j} = 0 \qquad \phi = 0$$

Moreover, if $R_{ij}\xi^i\xi^j$ is a negative definite form, then, from (2.73), we conclude

$$\xi_i = 0$$

and hence

THEOREM 2.14. In a compact orientable Riemannian

manifold V_n $(n \geq 2)$, there exists no vector field defining a conformal transformation which satisfies

$$R_{ij}\xi^i\xi^j \leq 0$$

unless we have

$$\xi_{i;j} = 0$$

and then automatically $R_{ij}\xi^i\xi^j = 0$.

Especially, if the manifold has negative definite Ricci curvature throughout, there exists no vector field defining an infinitesimal conformal transformation other than zero vector, and consequently there exists no one-parameter continuous group of conformal transformations. (Bochner [2], Yano [3]).

13. A NECESSARY AND SUFFICIENT CONDITION THAT A VECTOR BE A HARMONIC VECTOR

We know already that, if $\xi^i(x)$ is harmonic, that is, if

$$\xi_{i;j} - \xi_{j;i} = 0 \qquad \text{and} \qquad \xi^i{}_{;i} = 0$$

then also

$$g^{jk}\xi^i{}_{;j;k} - R^i{}_j\xi^j = 0 \tag{2.74}$$

and we are going to prove the converse.

For an arbitrary vector field $\xi^i(x)$ we put

$$\phi = \xi^i\xi_i$$

and form

$$\Delta\phi = 2[(g^{jk}\xi^i{}_{;j;k})\xi_i + \xi^{i;j}\xi_{i;j}]$$

and by Theorem 2.4, we find

$$\int_{V_n}[(g^{jk}\xi^i{}_{;j;k})\xi_i + \xi^{i;j}\xi_{i;j}]dv = 0 \tag{2.75}$$

On the other hand, we know that

$$\int_{V_n} [R_{ij}\xi^i\xi^j + \xi^{i;j}\xi_{j;i} - \xi^i{}_{;i}\xi^j{}_{;j}]dv = 0 \tag{2.76}$$

and thus, we find

$$\int_{V_n} [(g^{jk}\xi^i{}_{;j;k} - R^i{}_j\xi^j)\xi_i + \xi^{i;j}(\xi_{i;j} - \xi_{j;i}) + \xi^i{}_{;i}\xi^j{}_{;j}]dv = 0$$

which may be also written in the form

$$\int_{V_n} [(g^{jk}\xi^i{}_{;j;k} - R^i{}_j\xi^j)\xi_i + \frac{1}{2}(\xi^{i;j} - \xi^{j;i})(\xi_{i;j} - \xi_{j;i}) + \xi^i{}_{;i}\xi^j{}_{;j}]dv = 0 \tag{2.77}$$

This equation shows that, if the vector ξ^i satisfies (2.74), then we must have

$$\xi_{i;j} - \xi_{j;i} = 0 \qquad \text{and} \qquad \xi^i{}_{;i} = 0$$

that is to say, the vector must be harmonic. Thus we have

THEOREM 2.15. In a compact orientable Riemannian manifold V_n, a necessary and sufficient condition that a vector field $\xi^i(x)$ be a harmonic one is that it satisfy

$$g^{jk}\xi^i{}_{;j;k} - R^i{}_j\xi^j = 0$$

(de Rham [1]).

14. A NECESSARY AND SUFFICIENT CONDITION THAT A VECTOR BE A KILLING VECTOR

We know already that, if the vector $\xi^i(x)$ is a Killing vector, that is, if

$$\xi_{i;j} + \xi_{j;i} = 0$$

then we have

(2.78) $$g^{jk}\xi^i{}_{;j;k} + R^i{}_j\xi^j = 0\ ,\quad \xi^i{}_{;i} = 0$$

and we shall prove the converse.

Forming (2.75) + (2.76), we find

$$\int_{V_n} [(g^{jk}\xi^i{}_{;j;k} + R^i{}_j\xi^j)\xi_i + \xi^{i;j}(\xi_{i;j} + \xi_{j;i}) - \xi^i{}_{;i}\xi^j{}_{;j}]dv = 0$$

which may be written also in the form

(2.79) $$\int_{V_n} [(g^{jk}\xi^i{}_{;j;k} + R^i{}_j\xi^j)\xi_i + \frac{1}{2}(\xi^{i;j} + \xi^{j;i})(\xi_{i;j} + \xi_{j;i}) - \xi^i{}_{;i}\xi^j{}_{;j}]dv = 0$$

and this equation shows that, if the vector satisfies (2.78), then, we must have

$$\xi_{i;j} + \xi_{j;i} = 0$$

as claimed.

THEOREM 2.16. In a compact orientable Riemannian manifold V_n, a necessary and sufficient condition that a vector field $\xi^i(x)$ be a Killing vector is that it satisfy

$$g^{jk}\xi^i{}_{;j;k} + R^i{}_j\xi^j = 0 \qquad \xi^i{}_{;i} = 0$$

(Yano [3]).

15. MOTIONS AND AFFINE COLLINEATIONS

From the formula

$$L\,\Gamma^i_{jk} = \frac{1}{2} g^{ia}[(Lg_{aj})_{;k} + (Lg_{ak})_{;j} - (Lg_{jk})_{;a}]$$

it is evident that a motion in the Riemannian manifold V_n is necessarily an affine collineation in the manifold.

Conversely, if

$$\bar{x}^i = x^i + \xi^i(x)\delta t$$

is an infinitesimal affine collineation in V_n, then

$$\text{(2.80)} \qquad L\,\Gamma^i_{jk} = \xi^i{}_{;j;k} + R^i{}_{jkl}\xi^l = 0$$

and multiplying (2.80) by g^{jk} and contracting, we find

$$g^{jk}\xi^i{}_{;j;k} + R^i{}_l\xi^l = 0$$

On the other hand, on putting $i = j = a$ in (2.80) and adding with respect to a, we find

$$\xi^a{}_{;a;k} = 0$$

by virtue of the identity

$$R^a{}_{akl} = 0$$

and thus

$$\xi^a{}_{;a} = \text{const.}$$

But,

$$\int_{V_n} \xi^a{}_{;a}dv = 0$$

implies then

$$\xi^a{}_{;a} = 0$$

and thus, by Theorem 2.16, ξ^i is a Killing vector.

THEOREM 2.17. In a compact orientable Riemannian manifold V_n, an affine collineation is necessarily a motion. (Yano [3]).

CHAPTER III

HARMONIC AND KILLING TENSORS

1. SOME APPLICATIONS OF THE THEOREM OF HOPF-BOCHNER

In this section, we assume that the manifold is of class C^3 and g_{jk} of class C^2 .

We consider a tensor field $\xi_{i_1 i_2 \dots i_p}(x)$ of class C^2 and put

$$\Phi = \xi^{i_1 i_2 \dots i_p} \xi_{i_1 i_2 \dots i_p}$$

so that

$$\Delta\Phi = 2(\xi^{i_1 i_2 \dots i_p ; j} \xi_{i_1 i_2 \dots i_p ; j} + \xi^{i_1 i_2 \dots i_p} g^{jk} \xi_{i_1 i_2 \dots i_p ; j ; k})$$

Since

$$\xi^{i_1 i_2 \dots i_p ; j} \xi_{i_1 i_2 \dots i_p ; j}$$

is a positive definite form in

$$\xi_{i_1 i_2 \dots i_p ; j}$$

if $\xi_{i_1 i_2 \dots i_p}$ satisfies differential equations of the form

$$g^{jk} \xi_{i_1 i_2 \dots i_p ; j ; k} = T_{i_1 i_2 \dots i_p j_1 j_2 \dots j_p} \xi^{j_1 j_2 \dots j_p} \tag{3.1}$$

and if the quadratic form

$$T_{i_1 i_2 \dots i_p j_1 j_2 \dots j_p} \xi^{i_1 i_2 \dots i_p} \xi^{j_1 j_2 \dots j_p} \tag{3.2}$$

satisfies

$$T_{i_1 i_2 \cdots i_p j_1 j_2 \cdots j_p} \xi^{i_1 i_2 \cdots i_p} \xi^{j_1 j_2 \cdots j_p} \geq 0$$

then we have

$$\Delta \phi = 2 \left(\xi^{i_1 i_2 \cdots i_p ; j} \xi_{i_1 i_2 \cdots i_p ; j} + T_{i_1 i_2 \cdots i_p j_1 j_2 \cdots j_p} \xi^{i_1 i_2 \cdots i_p} \xi^{j_1 j_2 \cdots j_p} \right) \geq 0$$

and consequently, by Theorem 2.3, we have

$$\xi_{i_1 i_2 \cdots i_p ; j} = 0$$

and

$$T_{i_1 i_2 \cdots i_p j_1 j_2 \cdots j_p} \xi^{i_1 i_2 \cdots i_p} \xi^{j_1 j_2 \cdots j_p} = 0$$

If the quadratic form (3.2) is a positive definite form, then the latter relation implies

$$\xi_{i_1 i_2 \cdots i_p} = 0$$

and hence we have

THEOREM 3.1. In a compact Riemannian manifold V_n, there exists no tensor field which satisfies relations

$$g^{jk} \xi_{i_1 i_2 \cdots i_p ; j ; k} = T_{i_1 i_2 \cdots i_p j_1 j_2 \cdots j_p} \xi^{j_1 j_2 \cdots j_p}$$

and

$$T_{i_1 i_2 \cdots i_p j_1 j_2 \cdots j_p} \xi^{i_1 i_2 \cdots i_p} \xi^{j_1 j_2 \cdots j_p} \geq 0$$

unless we have

$$\xi_{i_1 i_2 \cdots i_p ; j} = 0$$

There are no such tensor fields other than the zero tensor if the quadratic form (3.2) is positive definite.

Now, if we take an arbitrary anti-symmetric tensor field

$$\xi_{i_1 i_2 \ldots i_p}(x)$$

of class C^2 in V_n, then, by using the Ricci identities suitably, we obtain

$$(3.3)\quad g^{jk}\xi_{i_1 i_2 \ldots i_p;j;k} - g^{jk}(\xi_{i_1 i_2 \ldots i_p;j} - \xi_{j i_2 \ldots i_p;i_1} - \cdots - \xi_{i_1 i_2 \ldots i_{p-1} j;i_p})_{;k}$$

$$-(\xi^a{}_{i_2 \ldots i_p;a;i_1} - \xi^a{}_{i_1 i_3 \ldots i_p;a;i_2} - \cdots - \xi^a{}_{i_2 \ldots i_{p-1} i_1;a;i_p})$$

$$= \sum_s^{1\ldots p} \xi_{i_1 \ldots i_{s-1} a i_{s+1} \ldots i_p} R^a{}_{i_s}$$

$$+ \sum_{s<t}^{1\ldots p} \xi_{i_1 \ldots i_{s-1} a i_{s+1} \ldots i_{t-1} b i_{t+1} \ldots i_p} R^{ab}{}_{i_s i_t}$$

Thus we can see that, if the anti-symmetric tensor $\xi_{i_1 i_2 \ldots i_p}$ satisfies

$$(3.4)\quad g^{jk}(\xi_{i_1 i_2 \ldots i_p;j} - \xi_{j i_2 \ldots i_p;i_1} - \cdots - \xi_{i_1 i_2 \ldots i_{p-1} j;i_p})_{;k}$$

$$+ (\xi^a{}_{i_2 \ldots i_p;a;i_1} - \xi^a{}_{i_1 i_3 \ldots i_p;a;i_2} - \cdots - \xi^a{}_{i_2 \ldots i_{p-1} i_1;a;i_p}) = 0$$

then it satisfies also

$$(3.5)\quad g^{jk}\xi_{i_1 i_2 \ldots i_p;j;k} = \sum_s^{1\ldots p} \xi_{i_1 \ldots i_{s-1} a i_{s+1} \ldots i_p} R^a{}_{i_s}$$

$$+ \sum_{s<t}^{1\ldots p} \xi_{i_1 \ldots i_{s-1} a i_{s+1} \ldots i_{t-1} b i_{t+1} \ldots i_p} R^{ab}{}_{i_s i_t}$$

and consequently

$$\xi^{i_1 i_2 \cdots i_p}(g^{jk}\xi_{i_1 i_2 \cdots i_p;j;k}) = pR_{ij}\xi^{ii_2 \cdots i_p}\xi^{j}{}_{i_2 \cdots i_p}$$

$$+ \frac{p(p-1)}{2} R_{ijkl}\xi^{iji_3 \cdots i_p}\xi^{kl}{}_{i_3 \cdots i_p}$$

If therefore we introduce the quadratic form

$$(3.6)\quad F(\xi_{i_1 i_2 \cdots i_p}) = R_{ij}\xi^{ii_2 \cdots i_p}\xi^{j}{}_{i_2 \cdots i_p} + \frac{p-1}{2} R_{ijkl}\xi^{iji_3 \cdots i_p}\xi^{kl}{}_{i_3 \cdots i_p}$$

which is quite meaningful, then we obtain the following theorem:

THEOREM 3.2. In a compact Riemannian manifold V_n, there exists no anti-symmetric tensor field which satisfies (3.4) and

$$F(\xi_{i_1 i_2 \cdots i_p}) \geq 0$$

unless we have

$$\xi_{i_1 i_2 \cdots i_p;j} = 0$$

and automatically

$$F(\xi_{i_1 i_2 \cdots i_p}) = 0$$

Especially, if the form

$$F(\xi_{i_1 i_2 \cdots i_p})$$

is positive definite, then there exists no anti-symmetric tensor field other than zero which satisfies (3.4).

Dually to (3.3) we also have

$$-pg^{jk}\xi_{i_1i_2\ldots i_p;j;k}+g^{jk}(p\xi_{i_1i_2\ldots i_p;j}+\xi_{ji_2\ldots i_p;i_1}+\cdots+\xi_{i_1i_2\ldots i_{p-1}j;i_p})_{;k}$$

$$-(\xi^a{}_{i_2\ldots i_p;a;i_1}-\xi^a{}_{i_1i_3\ldots i_p;a;i_2}-\cdots-\xi^a{}_{i_2\ldots i_{p-1}i_1;a;i_p})$$

$$=\sum_s^{1\ldots p}\xi_{i_1\ldots i_{s-1}ai_{s+1}\ldots i_p}R^a{}_{i_s}$$

$$+\sum_{s<t}^{1\ldots p}\xi_{i_1\ldots i_{s-1}ai_{s+1}\ldots i_{t-1}bi_{t+1}\ldots i_p}R^{ab}{}_{i_si_t}$$

Thus if the anti-symmetric tensor $\xi_{i_1i_2\ldots i_p}$ satisfies

$$(3.7)\quad g^{jk}(p\xi_{i_1i_2\ldots i_p;j}+\xi_{ji_2\ldots i_p;i_1}+\cdots+\xi_{i_1i_2\ldots i_{p-1}j;i_p})_{;k}$$

$$-(\xi^a{}_{i_2\ldots i_p;a;i_1}-\xi^a{}_{i_1i_3\ldots i_p;a;i_2}-\cdots-\xi^a{}_{i_2\ldots i_{p-1}i_1;a;i_p})=0$$

then it satisfies also

$$(3.8)\quad g^{jk}\xi_{i_1i_2\ldots i_p;j;k}+\frac{1}{p}\sum_s^{1\ldots p}\xi_{i_1\ldots i_{s-1}ai_{s+1}\ldots i_p}R^a{}_{i_s}$$

$$+\frac{1}{p}\sum_{s<t}^{1\ldots p}\xi_{i_1\ldots i_{s-1}ai_{s+1}\ldots i_{t-1}bi_{t+1}\ldots i_p}R^{ab}{}_{i_si_t}=0$$

and consequently we obtain the relation

$$\xi^{i_1i_2\ldots i_p}(g^{jk}\xi_{i_1i_2\ldots i_p;j;k})=-F\{\xi_{i_1i_2\ldots i_p}\}$$

with the minus sign, where the symbol $F\{\xi_{i_1i_2\ldots i_p}\}$ is defined as before, and hence we have

THEOREM 3.3. In a compact Riemannian manifold V_n, there exists no anti-symmetric tensor field which satisfies (3.7) and

$$F(\xi_{i_1 i_2 \cdots i_p}) \leq 0$$

unless we have

$$\xi_{i_1 i_2 \cdots i_p ; j} = 0$$

and then automatically

$$F(\xi_{i_1 i_2 \cdots i_p}) = 0$$

Especially, if the form

$$F(\xi_{i_1 i_2 \cdots i_p})$$

is negative definite, then there exists no anti-symmetric tensor field other than zero tensor field which satisfies (3.7).

2. HARMONIC TENSORS

Now, if

$$\xi_{i_1 i_2 \cdots i_p}$$

is a harmonic tensor field of order p, then it satisfies (2.58) and (2.59), and consequently it satisfies (3.4). Thus from Theorem 3.2, we have

THEOREM 3.4. In a compact Riemannian manifold V_n, there exists no harmonic tensor field of order p which satisfies

$$F(\xi_{i_1 i_2 \cdots i_p}) \geq 0$$

unless we have

$$\xi_{i_1 i_2 \cdots i_p ; j} = 0$$

and then automatically

$$F(\xi_{i_1 i_2 \cdots i_p}) = 0$$

Especially, if the form

$$F(\xi_{i_1 i_2 \ldots i_p})$$

is positive definite, then there exists no harmonic field of order p other than the zero tensor field, and consequently, if the manifold is orientable, we have $B_p = 0$ $(p = 1, 2, \ldots, n-1)$. (Lichnérowicz [1], Mogi [1], Tomonaga [1], Yano [4]).

3. KILLING TENSORS

For a Killing vector ξ_i and a geodesic $x^i(s)$ of the manifold, we have

$$\frac{\delta}{ds}\left(\xi_i \frac{dx^i}{ds}\right) = \frac{1}{2}(\xi_{i;j} + \xi_{j;i}) \frac{dx^i}{ds}\frac{dx^j}{ds} = 0$$

along the geodesic, and thus the length of the orthogonal projection of a Killing vector on the tangent of a geodesic is constant along the geodesic.

Conversely, if the length of the orthogonal projection of a vector field on the tangent of any geodesic is constant along this geodesic, then

$$\frac{1}{2}(\xi_{i;j} + \xi_{j;i}) \frac{dx^i}{ds}\frac{dx^j}{ds} = 0$$

implies

$$\xi_{i;j} + \xi_{j;i} = 0$$

Thus, a necessary and sufficient condition that a vector field ξ_i be a Killing vector is that the length of the orthogonal projection of the vector on the tangent of any geodesic be constant along the geodesic.

Next, for an anti-symmetric tensor field $\xi_{i_1 i_2 \ldots i_p}$ the quantity

$$\xi_{i i_2 \ldots i_p} \frac{dx^i}{ds}$$

is parallel along any geodesic $x^i(s)$, if and only if

$$\frac{\delta}{ds}\left(\xi_{i i_2 \ldots i_p} \frac{dx^i}{ds}\right) = \frac{1}{2}(\xi_{i i_2 \ldots i_p ; j} + \xi_{j i_2 \ldots i_p ; i}) \frac{dx^i}{ds}\frac{dx^j}{ds} = 0$$

that is, if and only if

$$\xi_{ii_2\ldots i_p;j} + \xi_{ji_2\ldots i_p;i} = 0 \tag{3.9}$$

and such an anti-symmetric tensor field

$$\xi_{i_1 i_2\ldots i_p}$$

we will call a Killing tensor. Equation (3.9) shows that the covariant derivative

$$\xi_{i_1 i_2\ldots i_p;j}$$

is not only anti-symmetric in $i_1, i_2, \ldots, i_p$ but also anti-symmetric in i_1 , and j . Thus we can see that

$$\xi_{i_1 i_2\ldots i_p;j}$$

is anti-symmetric in all its indices, and consequently, equation (3.9) is equivalent to

$$\xi_{i_1 i_2\ldots i_p;j} = \xi_{[i_1 i_2\ldots i_p;j]} \tag{3.10}$$

or explicitly

$$p\xi_{i_1 i_2\ldots i_p;j} + \xi_{ji_2\ldots i_p;i_1} + \cdots + \xi_{i_1 i_2\ldots i_{p-1}j;i_p} = 0 \tag{3.11}$$

If

$$\xi_{i_1 i_2\ldots i_p}$$

is a Killing tensor, then it is evident from (3.10) that $\xi_{i_1 i_2\ldots i_p}$ satisfies

$$\xi^{a}{}_{i_2\ldots i_p;a} = 0 \tag{3.12}$$

but (3.11) and (3.12) imply (3.7), and consequently (3.8), and thus, as a special case of Theorem 3.3, we have

THEOREM 3.5. In a compact Riemannian manifold V_n, there exists no Killing tensor field of order p which satisfies

$$F(\xi_{i_1 i_2 \dots i_p}) \leq 0$$

unless we have

$$\xi_{i_1 i_2 \dots i_p ; j} = 0$$

and then automatically

$$F(\xi_{i_1 i_2 \dots i_p}) = 0$$

Especially, if the form

$$F(\xi_{i_1 i_2 \dots i_p})$$

is negative definite, then there exists no Killing tensor of order p other than zero tensor. (Mogi [1], Yano [4]).

4. A FUNDAMENTAL FORMULA

In a compact orientable Riemannian manifold V_n, with an anti-symmetric tensor $\xi_{i_1 i_2 \dots i_p}$ we form

$$\xi^{i i_2 \dots i_p}{}_{;j} \xi^{j}{}_{i_2 \dots i_p}$$

and the divergence

$$(3.13) \qquad (\xi^{i i_2 \dots i_p}{}_{;j} \xi^{j}{}_{i_2 \dots i_p})_{;i} = \xi^{i i_2 \dots i_p}{}_{;j;i} \xi^{j}{}_{i_2 \dots i_p} + \xi^{i i_2 \dots i_p}{}_{;j} \xi^{j}{}_{i_2 \dots i_p ; i}$$

From the Ricci identity:

$$\xi^{ii_2\cdots i_p}{}_{;j;k} - \xi^{ii_2\cdots i_p}{}_{;k;j}$$

$$= \xi^{ai_2\cdots i_p}R^{i}{}_{ajk} + \xi^{iai_3\cdots i_p}R^{i_2}{}_{ajk} + \cdots + \xi^{ii_2\cdots i_{p-1}a}R^{i_p}{}_{ajk}$$

we have, by contracting with respect to i and k,

$$\xi^{ii_2\cdots i_p}{}_{;j;i} = \xi^{ii_2\cdots i_p}{}_{;i;j} + \xi^{ai_2\cdots i_p}R_{aj}$$

$$+ \xi^{iai_3\cdots i_p}R^{i_2}{}_{aji} + \cdots + \xi^{ii_2\cdots i_{p-1}a}R^{i_p}{}_{aji}$$

and consequently, on substituting this into (3.13), we obtain

$$(\xi^{ii_2\cdots i_p}{}_{;j}\xi^{j}{}_{i_2\cdots i_p})_{;i} = \xi^{ii_2\cdots i_p}{}_{;i;j}\xi^{j}{}_{i_2\cdots i_p} + R_{ij}\xi^{ii_2\cdots i_p}\xi^{j}{}_{i_2\cdots i_p}$$

$$+ (p-1)R_{ijkl}\xi^{iki_3\cdots i_p}\xi^{jl}{}_{i_3\cdots i_p} + \xi^{ii_2\cdots i_p}{}_{;j}\xi^{j}{}_{i_2\cdots i_p;i}$$

by virtue of

$$R_{ijkl} = R_{lkji}$$

But, according to the identity:

$$R_{ijkl} + R_{iklj} + R_{iljk} = 0$$

the term

$$(p-1)R_{ijkl}\xi^{iki_3\cdots i_p}\xi^{jl}{}_{i_3\cdots i_p}$$

appearing in the right-hand member of the above equation can also be written as

$$(p-1)R_{ijkl}\xi^{iki_3\cdots i_p}\xi^{jl}{}_{i_3\cdots i_p} = \frac{p-1}{2}R_{ijkl}\xi^{iji_3\cdots i_p}\xi^{kl}{}_{i_3\cdots i_p}$$

and thus we have

$$(3.14)\quad (\xi^{ii_2\cdots i_p}{}_{;j}\xi^{j}{}_{i_2\cdots i_p})_{;i} = \xi^{ii_2\cdots i_p}{}_{;i;j}\xi^{j}{}_{i_2\cdots i_p} + R_{ij}\xi^{ii_2\cdots i_p}\xi^{j}{}_{i_2\cdots i_p} + \frac{p-1}{2}R_{ijkl}\xi^{iji_3\cdots i_p}\xi^{kl}{}_{i_3\cdots i_p} + \xi^{ii_2\cdots i_p}{}_{;j}\xi^{j}{}_{i_2\cdots i_p;i}$$

Next, we consider

$$\xi^{ii_2\cdots i_p}{}_{;i}\xi^{j}{}_{i_2\cdots i_p}$$

and the divergence

$$(3.15)\quad (\xi^{ii_2\cdots i_p}{}_{;i}\xi^{j}{}_{i_2\cdots i_p})_{;j} = \xi^{ii_2\cdots i_p}{}_{;i;j}\xi^{j}{}_{i_2\cdots i_p} + \xi^{ii_2\cdots i_p}{}_{;i}\xi^{j}{}_{i_2\cdots i_p;j}$$

and from (3.14) - (3.15) we obtain

$$(3.16)\quad (\xi^{ii_2\cdots i_p}{}_{;j}\xi^{j}{}_{i_2\cdots i_p})_{;i} - (\xi^{ii_2\cdots i_p}{}_{;i}\xi^{j}{}_{i_2\cdots i_p})_{;j} = F(\xi_{i_1 i_2\cdots i_p}) + \xi^{ii_2\cdots i_p}{}_{;j}\xi^{j}{}_{i_2\cdots i_p;i} - \xi^{ii_2\cdots i_p}{}_{;i}\xi^{j}{}_{i_2\cdots i_p;j}$$

Integrating both members of (3.16) over the whole manifold, and applying Theorem 2.4, we get

$$(3.17)\quad \int_{V_n}(F(\xi_{i_1 i_2\cdots i_p}) + \xi^{ii_2\cdots i_p;j}\xi_{ji_2\cdots i_p;i} - \xi^{ii_2\cdots i_p}{}_{;i}\xi^{j}{}_{i_2\cdots i_p;j})dv = 0$$

where

$$\xi^{ii_2\cdots i_p;j} = \xi^{ii_2\cdots i_p}{}_{;a}g^{aj}$$

Now, the tensor

$$\xi^{ii_2\cdots i_p}$$

is anti-symmetric in all the indices, and hence

$$\xi^{ii_2\cdots i_p;j}\xi_{ji_2\cdots i_p;i} = \frac{1}{p}\,\xi^{i_1i_2\cdots i_p;j}\xi_{i_1i_2\cdots i_p;j} - \frac{p+1}{p}\,\xi^{[i_1i_2\cdots i_p;j]}\xi_{[i_1i_2\cdots i_p;j]}$$

where

$$\xi_{[i_1i_2\cdots i_p;j]}$$

denotes the anti-symmetric part of the tensor

$$\xi_{i_1i_2\cdots i_p;j}$$

and, on introducing this into (3.17) we obtain the relation:

$$(3.18)\qquad \int_{V_n}\Big(F(\xi_{i_1i_2\cdots i_p}) + \frac{1}{p}\,\xi^{i_1i_2\cdots i_p;j}\xi_{i_1i_2\cdots i_p;j} - \frac{p+1}{p}\,\xi^{[i_1i_2\cdots i_p;j]}\xi_{[i_1i_2\cdots i_p;j]} - \xi^{ii_2\cdots i_p}{}_{;i}\,\xi^{j}{}_{i_2\cdots i_p;j}\Big)dv = 0$$

which will be of equal importance. (Yano [4]).

5. SOME APPLICATIONS OF THE FUNDAMENTAL FORMULAS

If $\xi_{i_1i_2\cdots i_p}(x)$ is harmonic, then the substitution of

$$\xi_{[i_1i_2\cdots i_p;j]} = 0$$

and

$$\xi^{a}{}_{i_2\cdots i_p;a} = 0$$

in (3.18) gives

$$(3.19) \qquad \int_{V_n} (F(\xi_{i_1 i_2 \cdots i_p}) + \frac{1}{p} \xi^{i_1 i_2 \cdots i_p ; j} \xi_{i_1 i_2 \cdots i_p ; j}) dv = 0$$

and thus

$$F(\xi_{i_1 i_2 \cdots i_p}) \geq 0$$

implies

$$\xi_{i_1 i_2 \cdots i_p ; j} = 0$$

and

$$F(\xi_{i_1 i_2 \cdots i_p}) = 0$$

and

$$F(\xi_{i_1 i_2 \cdots i_p}) > 0$$

implies

$$\xi_{i_1 i_2 \cdots i_p} = 0$$

This gives another proof of Theorem 3.4 for an orientable manifold. (Yano [4]).

Similarly for a Killing tensor, (see (3.10) and (3.12)), we obtain

$$(3.20) \qquad \int_{V_n} (F(\xi_{i_1 i_2 \cdots i_p}) - \xi^{i_1 i_2 \cdots i_p ; j} \xi_{i_1 i_2 \cdots i_p ; j}) dv = 0$$

and thus

$$F(\xi_{i_1 i_2 \cdots i_p}) \leq 0$$

implies

$$\xi_{i_1 i_2 \cdots i_p ; j} = 0$$

and

$$F(\xi_{i_1 i_2 \ldots i_p}) = 0$$

and

$$F(\xi_{i_1 i_2 \ldots i_p}) < 0$$

implies

$$\xi_{i_1 i_2 \ldots i_p} = 0$$

This gives another proof of Theorem 3.5 for an orientable manifold. (Yano [4]).

6. CONFORMAL KILLING TENSOR

If a vector field ξ^i defines a one-parameter continuous group of conformal transformations, then

$$\xi_{i;j} + \xi_{j;i} = 2\phi g_{ij}$$

where $\phi = \frac{1}{n} \xi^i{}_{;i}$, and consequently we have

$$\frac{\delta}{ds} (\xi_i \frac{dx^i}{ds}) = \frac{1}{2} (\xi_{i;j} + \xi_{j;i}) \frac{dx^i}{ds} \frac{dx^j}{ds} = \phi$$

along any geodesic $x^i(s)$, and thus

$$\frac{\delta}{ds} (\xi_i \frac{dx^i}{ds})$$

depends only on the point and not on the direction of the geodesic passing through this point.

In order to obtain a corresponding property for an anti-symmetric tensor $\xi_{i_1 i_2 \ldots i_p}$, we assume that

$$\text{(3.21)} \qquad \frac{\delta}{ds} (\xi_{i i_2 \ldots i_p} \frac{dx^i}{ds}) = \frac{1}{2} (\xi_{i i_2 \ldots i_p ; j} + \xi_{j i_2 \ldots i_p ; i}) \frac{dx^i}{ds} \frac{dx^j}{ds}$$

depends only on the point and not on the direction of geodesic $x^i(s)$ passing through this point, and, from (3.21), we obtain

(3.22) $$\xi_{ii_2\ldots i_p;j} + \xi_{ji_2\ldots i_p;i} = 2\phi_{i_2\ldots i_p}g_{ij}$$

where

$$\phi_{i_2i_3\ldots i_p} = \frac{1}{n}g^{ij}\xi_{ii_2i_3\ldots i_p;j}$$

An anti-symmetric tensor field

$$\xi_{i_1i_2\ldots i_p}(x)$$

which satisfies equation (3.22) will be called a conformal Killing tensor field.

From (3.22), we have

$$\xi_{ji_2\ldots i_p;i} = -\xi_{ii_2\ldots i_p;j} + 2\phi_{i_2\ldots i_p}g_{ij}$$

and hence

$$\xi^{ii_2\ldots i_p;j}\xi_{ji_2\ldots i_p;i} = -\xi^{ii_2\ldots i_p;j}\xi_{ii_2\ldots i_p;j} + 2n\phi^{i_2\ldots i_p}\phi_{i_2\ldots i_p}$$

and therefore, by (3.17),

$$\int_{V_n}(F(\xi_{i_1i_2\ldots i_p}) - \xi^{ii_2\ldots i_p;j}\xi_{ii_2\ldots i_p;j} - n(n-2)\phi^{i_2\ldots i_p}\phi_{i_2\ldots i_p})dv = 0$$

Consequently we have

THEOREM 3.6. In a compact orientable Riemannian manifold V_n, there exists no conformal Killing tensor of order p which satisfies

$$F(\xi_{i_1i_2\ldots i_p}) \leq 0$$

unless we have

$$\xi_{i_1i_2\ldots i_p;j} = 0$$

and then automatically

$$F(\xi_{i_1 i_2 \dots i_p}) = 0$$

Especially, if the form

$$F(\xi_{i_1 i_2 \dots i_p})$$

is negative definite, then there exists no conformal Killing tensor of order p other than the zero tensor. (Yano [4]).

7. A NECESSARY AND SUFFICIENT CONDITION THAT AN ANTI-SYMMETRIC TENSOR BE A HARMONIC TENSOR OR A KILLING TENSOR

If we introduce the symbol

$$(3.23) \quad S(\xi_{i_1 i_2 \dots i_p}) = g^{jk}\xi_{i_1 i_2 \dots i_p;j;k} - \sum_s^{1\dots p} \xi_{i_1 \dots i_{s-1} a i_{s+1} \dots i_p} R^a{}_{i_s}$$
$$- \sum_{s<t}^{1\dots p} \xi_{i_1 \dots i_{s-1} a i_{s+1} \dots i_{t-1} b i_{t+1} \dots i_p} R^{ab}{}_{i_s i_t}$$

(frequently denoted by $\Delta\xi_{i_1 i_2 \dots i_p}$), then for a harmonic tensor we have

$$S(\xi_{i_1 i_2 \dots i_p}) = 0$$

(see (3.5)); and we are going to show the converse for a compact orientable Riemannian manifold.

For

$$\phi = \xi^{i_1 i_2 \dots i_p}\xi_{i_1 i_2 \dots i_p}$$

we have

$$\Delta\phi = 2[\xi^{i_1 i_2 \dots i_p}(g^{jk}\xi_{i_1 i_2 \dots i_p;j;k}) + \xi^{i_1 i_2 \dots i_p;j}\xi_{i_1 i_2 \dots i_p;j}]$$

and thus, integrating over the whole manifold, we find

$$(3.24)\qquad \int_{V_n} [\xi^{i_1 i_2 \cdots i_p}(g^{jk}\xi_{i_1 i_2 \cdots i_p;j;k}) + \xi^{i_1 i_2 \cdots i_p;j}\xi_{i_1 i_2 \cdots i_p;j}]dv = 0$$

Subtracting p times of (3.18) from (3.24), we have

$$(3.25)\qquad \int_{V_n} [\xi^{i_1 i_2 \cdots i_p} g^{jk}\xi_{i_1 i_2 \cdots i_p;j;k} - p\, F\{\xi_{i_1 i_2 \cdots i_p}\} + (p + 1)\xi^{[i_1 i_2 \cdots i_p;j]}\xi_{[i_1 i_2 \cdots i_p;j]} + p\xi^{i i_2 \cdots i_p}{}_{;i}\xi^{j}{}_{i_2 \cdots i_p;j}]dv = 0$$

which may be written also in the form

$$(3.26)\qquad \int_{V_n} [\xi^{i_1 i_2 \cdots i_p} S\{\xi_{i_1 i_2 \cdots i_p}\} + (p + 1)\xi^{[i_1 i_2 \cdots i_p;j]}\xi_{[i_1 i_2 \cdots i_p;j]} + p\xi^{i i_2 \cdots i_p}{}_{;i}\xi^{j}{}_{i_2 \cdots i_p;j}]dv = 0$$

and thus, if $S\{\xi_{i_1 i_2 \cdots i_p}\} = 0$ we must have

$$(3.27)\qquad \xi_{[i_1 i_2 \cdots i_p;j]} = 0 \quad \text{and} \quad \xi^{i}{}_{i_2 \cdots i_p;i} = 0$$

THEOREM 3.7. In a compact orientable Riemannian manifold V_n, a necessary and sufficient condition that an anti-symmetric tensor field

$$\xi_{i_1 i_2 \cdots i_p}$$

be harmonic is that it satisfy

$$S\{\xi_{i_1 i_2 \cdots i_p}\} = 0$$

(de Rham [1], Yano [4]).

For a Killing tensor the analogue to $S\{\xi_{i_1 i_2 \dots i_p}\}$ is the form

$$(3.28)\quad T\{\xi_{i_1 i_2 \dots i_p}\} = g^{jk}\xi_{i_1 i_2 \dots i_p;j;k} + \frac{1}{p}\sum_s^{1\dots p} \xi_{i_1 \dots i_{s-1} a i_{s+1} \dots i_p} R^a{}_{i_s} + \frac{1}{p}\sum_{s<t}^{1\dots p} \xi_{i_1 \dots i_{s-1} a i_{s+1} \dots i_{t-1} b i_{t+1} \dots i_p} R^{ab}{}_{i_s i_t}$$

(compare (3.8)); and we will again prove that, on a compact orientable Riemannian manifold,

$$T\{\xi_{i_1 i_2 \dots i_p}\} = 0$$

and

$$\xi^i{}_{i_2 \dots i_p;i} = 0$$

imply Killing property.

In fact, the analogue to (3.26) is

$$(3.29)\quad \int_{V_n} [\xi^{i_1 i_2 \dots i_p} T\{\xi_{i_1 i_2 \dots i_p}\} + \frac{p+1}{p}(\xi^{i_1 i_2 \dots i_p;j} - \xi^{[i_1 i_2 \dots i_p;j]})(\xi_{i_1 i_2 \dots i_p;j} - \xi_{[i_1 i_2 \dots i_p;j]}) - \xi^{i i_2 \dots i_p}{}_{;i}\xi^j{}_{i_2 \dots i_p;j}]dv = 0$$

and our conclusion follows.

> THEOREM 3.8. In a compact orientable Riemannian manifold V_n, a necessary and sufficient condition that an anti-symmetric tensor field be a Killing tensor is that it satisfy $T\{\xi_{i_1 i_2 \dots i_p}\} = 0$ and $\xi^i{}_{i_2 \dots i_p;i} = 0$. (Yano [4]).

CHAPTER IV

HARMONIC AND KILLING TENSORS IN FLAT MANIFOLDS

1. HARMONIC AND KILLING TENSORS IN A MANIFOLD OF CONSTANT CURVATURE

For

$$R_{ijkl} = \frac{R}{n(n-1)} (g_{jk}g_{il} - g_{jl}g_{ik})$$

and

$$R_{jk} = \frac{R}{n} g_{jk}$$

our form (3.6) has the value

$$F(\xi_{i_1 i_2 \cdots i_p}) = \frac{n-p}{n(n-1)} R\xi^{i_1 i_2 \cdots i_p} \xi_{i_1 i_2 \cdots i_p}$$

and hence we obtain the following conclusion.

In a compact Riemannian manifold V_n of positive constant curvature, there exists no harmonic tensor

$$\xi_{i_1 i_2 \cdots i_p}$$

other than zero, and consequently, in an orientable Riemannian manifold we have $B_p = 0$ for $p = 1, 2, \ldots, n-1$.

What is more important, if the constant curvature is negative (hyperbolic spaces), then there is no Killing tensor of order p $(p = 1, 2, \ldots, n-1)$.

And if $R = 0$, (compact flat manifold), there are $\binom{n}{p}$ independent harmonic tensors and Killing tensors.

The last statement follows from the fact that we may choose a coordinate system in which

$$\left\{ {i \atop jk} \right\} = 0$$

so that $\xi_{i_1 i_2 \cdots i_p ; j} = 0$ amounts to

$$\frac{\partial \xi_{i_1 i_2 \cdots i_p}}{\partial x^j} = 0$$

2. HARMONIC TENSORS AND KILLING TENSORS IN A CONFORMALLY FLAT MANIFOLD

A transformation of the metric

$$\bar{g}_{jk} = \rho^2 g_{jk} \tag{4.1}$$

where ρ is a scalar function, is called a conformal transformation of the Riemannian metric. Under a conformal transformation of metric, the curvature tensor $R^i{}_{jkl}$ will be transformed into

$$\bar{R}^i{}_{jkl} = R^i{}_{jkl} - \rho_{jk}\delta^i_l + \rho_{jl}\delta^i_k - g_{jk}\rho^i{}_l + g_{jl}\rho^i{}_k \tag{4.2}$$

where

$$\rho_{jk} = \rho_{j;k} - \rho_j \rho_k + \frac{1}{2} g^{bc}\rho_b \rho_c g_{jk} , \quad \rho^i{}_l = g^{ik}\rho_{kl} , \qquad \left(\rho_j = \frac{\partial \log \rho}{\partial x^j}\right)$$

but the tensor

$$C^i{}_{jkl} = R^i{}_{jkl} - \frac{1}{n-2}(R_{jk}\delta^i_l - R_{jl}\delta^i_k + g_{jk}R^i{}_l - g_{jl}R^i{}_k) + \frac{R}{(n-1)(n-2)}(g_{jk}\delta^i_l - g_{jl}\delta^i_k) \tag{4.3}$$

will be invariant.

Obviously if our Riemannian manifold can be reduced to a Euclidean manifold by a suitable conformal transformation, then $C^i{}_{jkl} = 0$, and J. A. Schouten has proved that for $n > 3$ this is also sufficient.

A manifold with $C^i{}_{jkl} = 0$ $(n > 3)$ will be called conformally flat.

For $C^i{}_{jkl} = 0$, we have

$$R_{ijkl} = \frac{1}{n-2}(R_{jk}g_{il} - R_{jl}g_{ik} + g_{jk}R_{il} - g_{jl}R_{ik})$$

$$- \frac{R}{(n-1)(n-2)}(g_{jk}g_{il} - g_{jl}g_{ik})$$

and if we substitute this into (3.6) we obtain

$$F(\xi_{i_1 i_2 \ldots i_p}) = \frac{n-2p}{n-2} R_{ij}\xi^{i i_2 \ldots i_p}\xi^{j}{}_{i_2 \ldots i_p}$$

$$+ \frac{p-1}{(n-1)(n-2)} R\,\xi^{i_1 i_2 \ldots i_p}\xi_{i_1 i_2 \ldots i_p}$$

Thus, if we assume that the quadratic form $R_{ij}\xi^i\xi^j$ is positive definite, and if we denote by L the smallest (positive) eigenvalue of the matrix $||R_{ij}||$, then we have

$$(4.4) \qquad R_{ij}\xi^i\xi^j \geq L\xi^i\xi_i \quad \text{and} \quad R = g^{ij}R_{ij} \geq nL > 0$$

and if we fix a point in the manifold and take a coordinate system in which $g_{ij} = \delta_{ij}$ at the fixed point, so that contravariant and covariant components of a tensor have the same values at this point, then, for $n \geq 2p$, we have

$$F(\xi_{i_1 i_2 \ldots i_p}) \geq \frac{n-2p}{n-2} L\xi^{i_1 i_2 \ldots i_p}\xi_{i_1 i_2 \ldots i_p} + \frac{n(p-1)}{(n-1)(n-2)} L\xi^{i_1 i_2 \ldots i_p}\xi_{i_1 i_2 \ldots i_p}$$

$$= \frac{n-p}{n-1} L\xi^{i_1 i_2 \ldots i_p}\xi_{i_1 i_2 \ldots i_p}$$

Thus the quadratic form

$$F(\xi_{i_1 i_2 \ldots i_p})$$

is positive definite, and this is so also in all coordinate systems, and by Theorem 3.4, we have $B_p = 0$ for $p = 1, 2, \ldots, [n/2]$. If we now apply Poincaré's duality theorem for Betti numbers we obtain

THEOREM 4.1. In a conformally flat compact orientable Riemannian manifold V_n, if the Ricci quadratic form $R_{ij}\xi^i\xi^j$ is positive definite, then we have $B_p = 0$, $(p = 1, 2, \ldots, n-1)$. (Bochner [5], Lichnérowicz [1]).

Next, if we assume that the Ricci quadratic form $R_{ij}\xi^i\xi^j$ is negative definite and denote by $-M$ the biggest (negative) eigenvalue of the matrix $||R_{ij}||$, then we have

$$R_{ij}\xi^i\xi^j \leq -M\xi^i\xi_i \quad \text{and} \quad R = g^{ij}R_{ij} \leq -nM < 0$$

and for $g_{ij} = \delta_{ij}$ we obtain for $n \geq 2p$,

$$\begin{aligned} F(\xi_{i_1 i_2 \ldots i_p}) &\leq -\frac{n-2p}{n-2} M\xi^{i_1 i_2 \ldots i_p}\xi_{i_1 i_2 \ldots i_p} \\ &\quad - \frac{n(p-1)}{(n-1)(n-2)} M\xi^{i_1 i_2 \ldots i_p}\xi_{i_1 i_2 \ldots i_p} \\ &= -\frac{n-p}{n-1} M\xi^{i_1 i_2 \ldots i_p}\xi_{i_1 i_2 \ldots i_p} \end{aligned}$$

and hence we have

THEOREM 4.2. In a conformally flat compact orientable Riemannian manifold V_n, if the Ricci quadratic form $R_{ij}\xi^i\xi^j$ is negative definite, then there exists no (conformal) Killing tensor field other than zero for $p = 1, 2, \ldots, [n/2]$.

CHAPTER V

DEVIATION FROM FLATNESS

1. DEVIATION FROM CONSTANCY OF CURVATURE

If

$$(5.1) \qquad R_{ijkl} = K(g_{jk}g_{il} - g_{jl}g_{ik}) , \qquad K > 0$$

then, for an anti-symmetric tensor ξ^{ij}, the quantity

$$- \frac{R_{ijkl}\xi^{ij}\xi^{kl}}{\xi^{ij}\xi_{ij}} = 2K$$

and is a positive constant.

Assume now more generally that we have

$$(5.2) \qquad 0 < A \leq - \frac{R_{ijkl}\xi^{ij}\xi^{kl}}{\xi^{ij}\xi_{ij}} \leq B$$

for every anti-symmetric tensor ξ^{ij}, A and B being constants.

If we put

$$\xi^{ij} = p^i q^j - p^j q^i$$

for two unit vectors p^i and q^i which are mutually orthogonal, then we get, from (5.2),

$$\frac{1}{2} A \leq - R_{ijkl}p^i q^j p^k q^l \leq \frac{1}{2} B$$

where $- R_{ijkl}p^i q^j p^k q^l$ is sectional curvature determined by a 2-plane containing p^i and q^i.

Now, for $n - 1$ unit vectors $q^i_{(1)}, q^i_{(2)}, \ldots, q^i_{(n-1)}$ which are orthogonal to p^i and to each other, we have

$$- R_{ijkl}p^ip^jp^kp^l = 0$$

$$\frac{1}{2} A \leq - R_{ijkl}p^iq^j_{(a)}p^kq^l_{(a)} \leq \frac{1}{2} B \quad (a = 1, 2, \ldots, n - 1)$$

and from this and

$$p^jp^l + \sum_{a=1}^{n-1} q^j_{(a)}q^l_{(a)} = g^{jl}$$

we obtain

$$(5.3) \qquad \frac{1}{2} (n - 1)A \leq R_{ik}p^ip^k \leq \frac{1}{2} (n - 1)B$$

From (5.3), we have

$$R_{ij}\xi^i\xi^j \geq \frac{1}{2} (n - 1)A\xi^i\xi_i$$

for any vector ξ^i and consequently

$$(5.4) \qquad R_{ij}\xi^{ia}\xi^j{}_a \geq \frac{1}{2} (n - 1)A\xi^{ij}\xi_{ij}$$

for any anti-symmetric tensor ξ^{ij} and, from (5.2), we have

$$(5.5) \qquad R_{ijkl}\xi^{ij}\xi^{kl} \geq - B\xi^{ij}\xi_{ij}$$

Thus, from (5.4) and (5.5), we find

$$R_{ij}\xi^{ia}\xi^j{}_a + \frac{p-1}{2} R_{ijkl}\xi^{ij}\xi^{kl} \geq \frac{1}{2} [(n - 1)A - (p - 1)B]\xi^{ij}\xi_{ij}$$

or

$$F(\xi_{i_1i_2\ldots i_p}) \geq \frac{1}{2} [(n - 1)A - (p - 1)B]\xi^{i_1i_2\ldots i_p}\xi_{i_1i_2\ldots i_p}$$

and for

$$(5.6) \qquad \frac{A}{B} > \frac{p-1}{n-1}$$

this is positive.

But, since we have

$$\frac{1}{2} > \frac{p-1}{n-1} \qquad \text{for } p = 1, 2, \ldots, [n/2]$$

we can say that, for

$$\frac{A}{B} = \frac{1}{2} \quad \text{or} \quad A = \frac{1}{2} B$$

the form

$$F(\xi_{i_1 i_2 \ldots i_p})$$

is positive definite for $p = 1, 2, \ldots, [n/2]$. Thus, applying Theorem 3.4, we have

THEOREM 5.1. In a compact orientable Riemannian manifold V_n , if the curvature tnesor satisfies

$$(5.7) \qquad 0 < \frac{1}{2} B \leq - \frac{R_{ijkl}\xi^{ij}\xi^{kl}}{\xi^{ij}\xi_{ij}} \leq B$$

for any anti-symmetric tensor ξ^{ij} , B being a constant, then all the Betti numbers B_p vanish, ($p = 1, 2, \ldots, n - 1$) . (Bochner and Yano [1]).

This result may be compared with a recent result of H. E. Rauch [1].

Next, if we assume that

$$(5.8) \qquad - A \leq - \frac{R_{ijkl}\xi^{ij}\xi^{kl}}{\xi^{ij}\xi_{ij}} \leq - B < 0$$

for any anti-symmetric tensor ξ^{ij} , then

$$F(\xi_{i_1 i_2 \ldots i_p}) \leq \frac{1}{2} [(p - 1)A - (n - 1)B] \xi^{i_1 i_2 \ldots i_p} \xi_{i_1 i_2 \ldots i_p}$$

and for

$$\frac{p-1}{n-1} < \frac{B}{A}$$

this will be negative.

Thus, applying Theorem 3.5, we have

THEOREM 5.2. In a compact Riemannian manifold, if the curvature tensor satisfies

$$- A \leq - \frac{R_{ijkl}\xi^{ij}\xi^{kl}}{\xi^{ij}\xi_{ij}} \leq - \frac{1}{2} A < 0 \tag{5.9}$$

for any anti-symmetric tensor ξ^{ij} , A being a constant, then there exists no Killing tensor of order p where p = 1, 2, ..., [n/2] .

2. DEVIATION FROM PROJECTIVE FLATNESS

Consider an n-dimensional Riemannian manifold. If there exists, for any coordinate neighborhood of the manifold, a one-to-one correspondence between this neighborhood and a domain in Euclidean space such that any geodesic of the Riemannian manifold corresponds to a straight line in the Euclidean space, then we say that the Riemannian manifold is locally projectively flat.

For $n \geq 3$, a necessary and sufficient condition that the manifold be locally projectively flat is that the so-called Weyl projective curvature tensor W_{ijkl} vanish, where

$$W_{ijkl} = R_{ijkl} - \frac{1}{n-1}(R_{jk}g_{il} - R_{jl}g_{ik}) \tag{5.10}$$

Now,

$$W_{ijkl} \equiv R_{ijkl} - \frac{1}{n-1}(R_{jk}g_{il} - R_{jl}g_{ik}) = 0 \tag{5.11}$$

implies

$$R_{jk}g_{il} - R_{jl}g_{ik} + R_{ik}g_{jl} - R_{il}g_{jk} = 0$$

and hence

$$R_{jk} = \frac{1}{n} Rg_{jk}$$

and substituting this into (5.11), we find

$$R_{ijkl} = \frac{R}{n(n-1)}(g_{jk}g_{il} - g_{jl}g_{ik}) \tag{5.12}$$

and this shows that the manifold is also of constant curvature.

Conversely, if the manifold is of constant curvature, then its Riemannian-Christoffel curvature tensor has the form (5.12) and its Ricci tensor has the form $R_{ij} = (R/n)g_{ij}$, and consequently, it will be easily verified that $W_{ijkl} = 0$, that is to say, that the manifold is locally projectively flat.

If we substitute

$$R_{ijkl} = W_{ijkl} + \frac{1}{n-1}(R_{jk}g_{il} - R_{jl}g_{ik})$$

into (3.6), we find

$$(5.13) \quad F(\xi_{i_1 i_2 \cdots i_p}) = \frac{n-p}{n-1} R_{ij}\xi^{i i_2 \cdots i_p}\xi^{j}{}_{i_2 \cdots i_p} + \frac{p-1}{2} W_{ijkl}\xi^{iji_3 \cdots i_p}\xi^{kl}{}_{i_3 \cdots i_p}$$

and in order to measure deviation from projective flatness, we introduce the quantity

$$(5.14) \qquad W = \sup_\xi \frac{|W_{ijkl}\xi^{ij}\xi^{kl}|}{\xi^{ij}\xi_{ij}}, \qquad (\xi^{ij} = -\xi^{ji})$$

Now, if we assume that $R_{ij}\xi^i\xi^j$ is positive definite and denote by L the smallest (positive) eigenvalue of the matrix $||R_{ij}||$, then we have

$$R_{ij}\xi^i\xi^j \geq L\xi^i\xi_i \quad \text{and} \quad R = g^{ij}R_{ij} \geq nL > 0$$

and thus, for $g_{ij} = \delta_{ij}$, we have

$$R_{ij}\xi^{ii_2 \cdots i_p}\xi^{j}{}_{i_2 \cdots i_p} \geq L\xi^{i_1 i_2 \cdots i_p}\xi_{i_1 i_2 \cdots i_p}$$

$$W_{ijkl}\xi^{iji_3 \cdots i_p}\xi^{kl}{}_{i_3 \cdots i_p} \geq -W\xi^{i_1 i_2 \cdots i_p}\xi_{i_1 i_2 \cdots i_p}$$

Consequently, we have, from (5.13),

$$F(\xi_{i_1 i_2 \cdots i_p}) \geq \left(\frac{n-p}{n-1} L - \frac{p-1}{2} W\right)\xi^{i_1 i_2 \cdots i_p}\xi_{i_1 i_2 \cdots i_p}$$

and we obtain the following conclusion:

THEOREM 5.3. In a compact orientable Riemannian manifold V_n which has a positive Ricci curvature, if

$$\frac{n-p}{n-1} L > \frac{p-1}{2} W \tag{5.15}$$

then there exists no harmonic tensor of order p other than the zero tensor, and consequently $B_p = 0$, $(p = 1, 2, \ldots, n-1)$. (Bochner [5], Yano [4]).

Similarly if $R_{ij}\xi^i\xi^j$ is negative definite and if we denote by $-M$ the biggest (negative) eigenvalue of the matrix $||R_{ij}||$ then

$$F(\xi_{i_1 i_2 \ldots i_p}) \leq \left(-\frac{n-p}{n-1} M + \frac{p-1}{2} W\right)\xi^{i_1 i_2 \ldots i_p}\xi_{i_1 i_2 \ldots i_p}$$

and from Theorem 3.5 and 3.6 we have

THEOREM 5.4. In a compact Riemannian manifold V_n which has a negative Ricci curvature, if

$$\frac{n-p}{n-1} M > \frac{p-1}{2} W \tag{5.16}$$

then there exists no (conformal) Killing tensor of order p other than the zero tensor, $(p = 1, 2, \ldots, n-1)$.

3. DEVIATION FROM CONCIRCULAR FLATNESS

In a Riemannian manifold, a geodesic circle is defined as a curve whose first curvature is constant and whose second, third, ..., curvatures are all zero. The differential equations of a geodesic circle are given by

$$\frac{\delta^3 x^i}{ds^3} + \frac{dx^i}{ds} g_{jk} \frac{\delta^2 x^j}{ds^2}\frac{\delta^2 x^k}{ds^2} = 0 \tag{5.17}$$

where δ/ds denotes the covariant differentiation along the curve, s being the arc length.

Now, under an arbitrary conformal transformation

(5.18) $$\bar{g}_{jk} = \rho^2 g_{jk}$$

any geodesic circle will be transformed into a geodesic circle if and only if the function ρ satisfies the relation

(5.19) $$\rho_{j;k} - \rho_j\rho_k = \phi g_{jk}$$

where

$$\rho_j = \frac{\partial \log\rho}{\partial x^j}$$

and such a conformal transformation will be called concircular, (Yano [1]).

The tensor

(5.20) $$Z^i{}_{jkl} = R^i{}_{jkl} - \frac{R}{n(n-1)}(g_{jk}\delta^i_l - g_{jl}\delta^i_k)$$

is invariant under any concircular transformation, and

(5.21) $$Z^i{}_{jkl} = 0$$

is a necessary and sufficient condition that a Riemannian manifold be reducible to a Euclidean space by a suitable concircular transformation.

We shall call such a Riemannian manifold concircularly flat. It is easily seen that if the manifold is concircularly flat, then it is of constant curvature, and conversely if the manifold is of constant curvature, then it is concircularly flat.

If we substitute

$$R_{ijkl} = Z_{ijkl} + \frac{R}{n(n-1)}(g_{jk}g_{il} - g_{jl}g_{ik})$$

into (3.7), we obtain

(5.22) $$F(\xi_{i_1 i_2 \cdots i_p}) = R_{ij}\xi^{ii_2\cdots i_p}\xi^j{}_{i_2\cdots i_p} - \frac{p-1}{n(n-1)}R\xi^{i_1 i_2\cdots i_p}\xi_{i_1 i_2\cdots i_p}$$
$$+ \frac{p-1}{2}Z_{ijkl}\xi^{iji_3\cdots i_p}\xi^{kl}{}_{i_3\cdots i_p}$$

and in order to measure the deviation from concircular flatness, we introduce the quantity

$$(5.23) \qquad Z = \sup_\xi \frac{|Z_{ijkl}\xi^{ij}\xi^{kl}|}{\xi^{ij}\xi_{ij}}, \qquad (\xi^{ij} = -\xi^{ji})$$

If $R_{ij}\xi^i\xi^j$ is positive definite, we have

$$F(\xi_{i_1 i_2 \cdots i_p}) \geq (L - \frac{p-1}{n(n-1)} R - \frac{p-1}{2} Z)\xi^{i_1 i_2 \cdots i_p}\xi_{i_1 i_2 \cdots i_p}$$

and if $R_{ij}\xi^i\xi^j$ is negative definite, we have

$$F(\xi_{i_1 i_2 \cdots i_p}) \leq (-M - \frac{p-1}{n(n-1)} R + \frac{p-1}{2} Z)\xi^{i_1 i_2 \cdots i_p}\xi_{i_1 i_2 \cdots i_p}$$

and hence the following conclusion.

THEOREM 5.5. In a compact orientable Riemannian manifold V_n, for positive Ricci curvature, if

$$L - \frac{p-1}{n(n-1)} R > \frac{p-1}{2} Z$$

then $B_p = 0$, $p = 1, 2, \ldots, n-1$, and for negative Ricci curvature, if

$$M + \frac{p-1}{n(n-1)} R > \frac{p-1}{2} Z$$

then there exists no (conformal) Killing tensor of order p other than zero, $p = 1, 2, \ldots, n-1$. (Yano [4]).

4. DEVIATION FROM CONFORMAL FLATNESS

Returning to the Weyl conformal curvature tensor, if we substitute from (4.3) into (3.6), we find

$$(5.24) \qquad F(\xi_{i_1 i_2 \cdots i_p}) = \frac{n-2p}{n-2} R_{ij}\xi^{i i_2 \cdots i_p}\xi^j{}_{i_2 \cdots i_p}$$

$$+ \frac{(p-1)R}{(n-1)(n-2)} \xi^{i_1 i_2 \cdots i_p}\xi_{i_1 i_2 \cdots i_p} + \frac{p-1}{2} C_{ijkl}\xi^{iji_3 \cdots i_p}\xi^{kl}{}_{i_3 \cdots i_p}$$

In order to measure the deviation from conformal flatness we introduce the quantity

$$C = \sup_\xi \frac{|C_{ijkl}\xi^{ij}\xi^{kl}|}{\xi^{ij}\xi_{ij}}, \qquad (\xi^{ij} = -\xi^{ji}) \tag{5.25}$$

If $R_{ij}\xi^i\xi^j$ is positive definite, then we have, for $n \geq 2p$,

$$F(\xi_{i_1 i_2 \ldots i_p}) \geq \left(\frac{n-2p}{n-2} L + \frac{p-1}{(n-1)(n-2)} R - \frac{p-1}{2} C\right)\xi^{i_1 i_2 \ldots i_p}\xi_{i_1 i_2 \ldots i_p}$$

and if $R_{ij}\xi^i\xi^j$ is negative definite, we have

$$F(\xi_{i_1 i_2 \ldots i_p}) \leq \left(-\frac{n-2p}{n-2} M + \frac{p-1}{(n-1)(n-2)} R + \frac{p-1}{2} C\right)\xi^{i_1 i_2 \ldots i_p}\xi_{i_1 i_2 \ldots i_p}$$

hence the following conclusion.

THEOREM 5.6. For positive Ricci curvature, if

$$\frac{n-2p}{n-2} L + \frac{p-1}{(n-1)(n-2)} R > \frac{p-1}{2} C$$

then $B_p = 0$, $p = 1, 2, \ldots, n-1$, and for negative Ricci curvature, if

$$\frac{n-2p}{n-2} M - \frac{p-1}{(n-1)(n-2)} R > \frac{p-1}{2} C \tag{5.26}$$

then there is no (conformal) Killing tensor of order p other than zero, $p = 1, 2, \ldots, [n/2]$. (Bochner [5], Mogi [1], Yano [4]).

CHAPTER VI

SEMI-SIMPLE GROUP SPACES

1. SEMI-SIMPLE GROUP SPACES

Take a compact semi-simple group space with Maurer-Cartan equations

$$(6.1)\qquad h^l_b \frac{\partial h^i_c}{\partial x^l} - h^l_c \frac{\partial h^i_b}{\partial x^l} = c_{bc}{}^a h^i_a \qquad (a, b, c, \ldots = \dot{1}, \dot{2}, \ldots, \dot{n})$$

where $c_{bc}{}^a$ are the constants of structure, (Eisenhart [2]).

If we put

$$(6.2)\qquad g_{bc} = - c_{be}{}^f c_{cf}{}^e$$

then, for a semi-simple group the rank of the matrix $||g_{bc}||$ is n and, since the space is compact, the quadratic form $g_{bc}z^b z^c$ is positive definite. Thus, denoting by $||g^{ab}||$ the inverse of the matrix $||g_{bc}||$, we can use g^{ab} and g_{bc} to raise and lower the indices $a, b, c, \ldots, f$.

Thus multiplying the Jacobi identity:

$$c_{ab}{}^e c_{ce}{}^f + c_{bc}{}^e c_{ae}{}^f + c_{ca}{}^e c_{be}{}^f = 0$$

by $c_{df}{}^a$ and contracting over a and f, we get

$$c_{bcd} \equiv c_{bc}{}^e g_{ed} = - c_{ab}{}^e c_{ce}{}^f c_{df}{}^a + c_{ab}{}^e c_{de}{}^f c_{cf}{}^a$$

which shows that c_{bcd} is anti-symmetric in all the indices b, c, and d.

If we put

$$(6.3)\qquad g^{ij} = h^i_a h^j_b g^{ab}$$

and denote by $||g_{jk}||$ the inverse matrix of $||g^{ij}||$, then we have

$$g_{jk} = h_j^b h_k^c g_{bc} \tag{6.4}$$

where

$$h_j^b = g^{bc} g_{jk} h_c^k \tag{6.5}$$

and the quadratic differential form

$$ds^2 = g_{jk} dx^j dx^k \tag{6.6}$$

is positive definite. We give this metric to our semi-simple group space.

As we have $h_a^i h_j^a = \delta_j^i$, we have, from (6.1),

$$\Omega_{jk}{}^i = \frac{1}{2} h_a^i \left(\frac{\partial h_j^a}{\partial x^k} - \frac{\partial h_k^a}{\partial x^j} \right) \tag{6.7}$$

where we have put

$$\Omega_{jk}{}^i = \frac{1}{2} c_{bc}{}^a h_j^b h_k^c h_a^i \tag{6.8}$$

$\Omega_{jk}{}^i$ being a tensor whose covariant components Ω_{jki} are anti-symmetric in all the indices.

Now, taking account of (6.3) and (6.4), we calculate the Christoffel symbols $\{{}^{\,i}_{jk}\}$ formed with g_{jk}. By a straightforward calculation, we find

$$\{{}^{\,i}_{jk}\} = \frac{1}{2} h_a^i \left(\frac{\partial h_j^a}{\partial x^k} + \frac{\partial h_k^a}{\partial x^j} \right) \tag{6.9}$$

Thus, denoting by a semi-colon the covariant derivative with respect to $\{{}^{\,i}_{jk}\}$, we find

$$h_{j;k}^a = \frac{1}{2} \left(\frac{\partial h_j^a}{\partial x^k} - \frac{\partial h_k^a}{\partial x^j} \right) = \Omega_{jk}{}^i h_i^a \tag{6.10}$$

and consequently

$$h_{a;k}^i = \Omega^i{}_{kl} h_a^l = \Omega_{kl}{}^i h_a^l \tag{6.11}$$

Equations (6.10) show that we have $h^a_{j;k} + h^a_{k;j} = 0$, and consequently the vectors h^a_i define translations in our space.

Thus, we have, from (6.8),

$$\Omega_{jk}{}^i{}_{;l} = \Omega_{jk}{}^p\Omega_{lp}{}^i + \Omega_{kl}{}^p\Omega_{jp}{}^i + \Omega_{lj}{}^p\Omega_{kp}{}^i = 0 \tag{6.12}$$

by virtue of the Jacobi identity.

Now, if we put

$$E^i_{jk} = h^i_a \frac{\partial h^a_j}{\partial x^k}$$

we have, from (6.7) and (6.9),

$$\left\{ {i \atop jk} \right\} = \frac{1}{2}(E^i_{jk} + E^i_{kj}), \qquad \Omega_{jk}{}^i = \frac{1}{2}(E^i_{jk} - E^i_{kj})$$

from which

$$E^i_{jk} = \left\{ {i \atop jk} \right\} + \Omega_{jk}{}^i$$

The curvature tensor formed with the affine connection E^i_{jk} being zero, we have

$$0 = R^i{}_{jkl} + \Omega_{jk}{}^i{}_{;l} - \Omega_{jl}{}^i{}_{;k} + \Omega_{jk}{}^s\Omega_{sl}{}^i - \Omega_{jl}{}^s\Omega_{sk}{}^i$$

from which, by virtue of (6.12) and of the Jacobi identity, we find

$$R^i{}_{jkl} = \Omega_{kl}{}^s\Omega_{sj}{}^i$$

or

$$R_{ijkl} = -\Omega_{ijs}\Omega_{kl}{}^s \tag{6.13}$$

Multiplying this equation by g^{il} and contracting over i and l, we find

$$R_{jk} = \frac{1}{4} g_{jk} \tag{6.14}$$

by virtue of

(6.15) $$-\Omega^{r}{}_{js}\Omega_{kr}{}^{s} = -\Omega_{js}{}^{r}\Omega_{kr}{}^{s} = \frac{1}{4}g_{jk}$$

Thus, our space is an Einstein space with positive scalar curvature.

2. A THEOREM ON CURVATURE OF A SEMI-SIMPLE GROUP SPACE

Now, we shall prove the following

THEOREM 6.1. In a semi-simple group space with the metric tensor (6.4), we have

$$0 \geq R_{ijkl}\xi^{ij}\xi^{kl} \geq -\frac{1}{4}\xi^{ij}\xi_{ij}$$

for any $\xi^{ij} = -\xi^{ji}$.

To prove this we fix a point in the space and take a coordinate system in which $g_{ij} = \delta_{ij}$ at this point, and write all indices as subscripts.

We have, from (6.15),

$$\sum_{i,j}^{1\dots n} 4\,\Omega_{ijs}\Omega_{ijt} = \delta_{st}$$

or

$$\sum_{i<j}^{1\dots n}(2\sqrt{2}\,\Omega_{ijs})(2\sqrt{2}\,\Omega_{ijt}) = \delta_{st}$$

Consequently, $2\sqrt{2}\,\Omega_{ijs}$ $(i < j;\ s = 1, 2, \dots, n)$ represent n unit vectors orthogonal to each other in $(1/2)n(n-1)$-dimensional Euclidean space. Thus, if we denote by Ω_{ijA} $(i < j;\ A = n+1, \dots, (1/2)n(n-1))$ the $(1/2)n(n-1)-n$ unit vectors orthogonal to each other and also to $2\sqrt{2}\,\Omega_{ijs}$, then we have

$$\sum_{s=1}^{n}(2\sqrt{2}\,\Omega_{ijs})(2\sqrt{2}\,\Omega_{kls}) + \sum_{A=n+1}^{\frac{1}{2}n(n-1)}\Omega_{ijA}\Omega_{klA} = \delta_{(ij)(kl)}$$

$$(i < j;\ k < l)$$

from which

$$8\sum_{s=1}^{n}\Big(\sum_{i<j}^{1\cdots n}\Omega_{ijs}\xi_{ij}\Big)^2 + \sum_{A=n+1}^{\frac{1}{2}n(n-1)}\Big(\sum_{i<j}^{1\cdots n}\Omega_{ijA}\xi_{ij}\Big)^2 = \sum_{i<j}^{1\cdots n}(\xi_{ij})^2$$

and consequently

$$\sum_{s=1}^{n}\sum_{i,j}^{1\cdots n}\sum_{k,l}^{1\cdots n}\Omega_{ijs}\Omega_{kls}\xi_{ij}\xi_{kl} \leq \frac{1}{4}\sum_{i,j}^{1\cdots n}\xi_{ij}\xi_{ij}$$

Thus, we have proved the tensor inequality

$$\Omega_{ijs}\Omega_{kl}{}^{s}\xi^{ij}\xi^{kl} \leq \frac{1}{4}\xi^{ij}\xi_{ij}$$

at a fixed point in a special coordinate system, and therefore this holds in general. (Yano [4]).

3. HARMONIC TENSORS IN A SEMI-SIMPLE GROUP SPACE

We now assume that there exists a harmonic tensor field

$$\xi_{i_1 i_2 \cdots i_p}$$

in our semi-simple group space, and then formula (3.18) gives

$$\int (F\{\xi_{i_1 i_2 \cdots i_p}\} + \frac{1}{p}\xi^{i_1 i_2 \cdots i_p ; j}\xi_{i_1 i_2 \cdots i_p ; j})dv = 0$$

For $p = 1$, we have

$$\int (\frac{1}{4}\xi^{i}\xi_{i} + \xi^{i;j}\xi_{i;j})dv = 0$$

which shows that $\xi_i = 0$, and hence we have $B_1 = 0$.

For $p = 2$, we have

$$\int (\frac{1}{4}\xi^{i_1 i_2}\xi_{i_1 i_2} + \frac{1}{2}R_{ijkl}\xi^{ij}\xi^{kl} + \frac{1}{2}\xi^{i_1 i_2 ; j}\xi_{i_1 i_2 ; j})dv = 0$$

but we have, by Theorem 6.1,

$$\frac{1}{4}\xi^{i_1i_2}\xi_{i_1i_2} + \frac{1}{2}R_{ijkl}\xi^{ij}\xi^{kl} \geq \frac{1}{4}\xi^{i_1i_2}\xi_{i_1i_2} - \frac{1}{2}\frac{1}{4}\xi^{i_1i_2}\xi_{i_1i_2} = \frac{1}{8}\xi^{i_1i_2}\xi_{i_1i_2}$$

Thus we must have $\xi_{ij} = 0$, and hence we also have $B_2 = 0$.

For $p = 3$, we have

$$\int(\frac{1}{4}\xi^{i_1i_2i_3}\xi_{i_1i_2i_3} + R_{ijkl}\xi^{iji_3}\xi^{kl}{}_{i_3} + \frac{1}{3}\xi^{i_1i_2i_3;j}\xi_{i_1i_2i_3;j})dv = 0$$

But, if we fix a point in the space and choose a coordinate system in which $g_{ij} = \delta_{ji}$ at this point, then we have, by Theorem 6.1,

$$\frac{1}{4}\xi^{i_1i_2i_3}\xi_{i_1i_2i_3} + R_{ijkl}\xi^{iji_3}\xi^{kl}{}_{i_3} \geq \frac{1}{4}\xi^{i_1i_2i_3}\xi_{i_1i_2i_3} - \frac{1}{4}\xi^{i_1i_2i_3}\xi_{i_1i_2i_3} = 0$$

and thus, we must have $\xi_{i_1i_2i_3;j} = 0$.

THEOREM 6.2. In a compact semi-simple group space, we have $B_1 = B_2 = 0$ as well known.

Also a harmonic tensor of the third order must have vanishing covariant derivative. But the tensor Ω_{ijk} is such a tensor which is not identically zero and hence we have $B_3 \geq 1$.

4. DEVIATION FROM FLATNESS

Now, in our semi-simple group space, we have

$$W_{ijkl} = Z_{ijkl} = C_{ijkl} = R_{ijkl} - \frac{1}{4(n-1)}(g_{jk}g_{il} - g_{jl}g_{ik})$$

and consequently

$$W_{ijkl}\xi^{ij}\xi^{kl} = Z_{ijkl}\xi^{ij}\xi^{kl} = C_{ijkl}\xi^{ij}\xi^{kl} = R_{ijkl}\xi^{ij}\xi^{kl} + \frac{1}{2(n-1)}\xi^{ij}\xi_{ij}$$

and hence, by Theorem 6.1,

$$\frac{1}{2(n-1)}\xi^{ij}\xi_{ij} \geq W_{ijkl}\xi^{ij}\xi^{kl} = Z_{ijkl}\xi^{ij}\xi^{kl} = C_{ijkl}\xi^{ij}\xi^{kl} \geq -\frac{n-3}{4(n-1)}\xi^{ij}\xi_{ij}$$

THEOREM 6.3. In a semi-simple group space, we have

$$\frac{1}{2(n-1)} \geq \frac{W_{ijkl}\xi^{ij}\xi^{kl}}{\xi^{ij}\xi_{ij}} = \frac{Z_{ijkl}\xi^{ij}\xi^{kl}}{\xi^{ij}\xi_{ij}} = \frac{C_{ijkl}\xi^{ij}\xi^{kl}}{\xi^{ij}\xi_{ij}} \geq -\frac{n-3}{4(n-1)}$$

and consequently we have

$$W = Z = C = \begin{cases} \dfrac{1}{2(n-1)} & (n \leq 5) \\ \\ \dfrac{n-3}{4(n-1)} & (n \geq 5) \end{cases}$$

(Bochner [5], Yano [4]).

CHAPTER VII

PSEUDO-HARMONIC TENSORS AND PSEUDO-KILLING TENSORS IN METRIC MANIFOLDS WITH TORSION

1. METRIC MANIFOLDS WITH TORSION

We consider an n-dimensional compact manifold V_n on which there is given a positive definite metric

$$ds^2 = g_{jk}dx^j dx^k \tag{7.1}$$

and a metric connection E^i_{jk}, so that

$$g_{jk|l} \equiv \frac{\partial g_{jk}}{\partial x^l} - g_{sk}E^s_{jl} - g_{js}E^s_{kl} = 0 \tag{7.2}$$

where the solidus denotes covariant differentiation with respect to E^i_{jk}.

The connection E^i_{jk} needs not be symmetric, $E^i_{jk} \neq E^i_{kj}$, and the entity

$$S_{jk}{}^i = \frac{1}{2}(E^i_{jk} - E^i_{kj}) \tag{7.3}$$

will be called the torsion tensor. We define g^{is} by $g^{is}g_{sj} = \delta^i_j$, and we will use this to pull indices up and down, so that for instance

$$S^i{}_{jk} = g^{is}g_{kt}S_{sj}{}^t \tag{7.4}$$

and we note that, in virtue of (7.2), the pulling up and down of indices is commutative with covariant differentiation.

From (7.2), we have

$$\frac{\partial g_{sj}}{\partial x^k} - g_{tj}E^t_{sk} - g_{st}E^t_{jk} = 0$$

$$\frac{\partial g_{sk}}{\partial x^j} - g_{tk}E^t_{sj} - g_{st}E^t_{kj} = 0$$

$$-\frac{\partial g_{jk}}{\partial x^s} + g_{tk}E^t_{js} + g_{jt}E^t_{ks} = 0$$

and on multiplying the sum of these equations by $(1/2)g^{is}$ and contracting over s we find

$$(7.5) \qquad E^i_{jk} = \{^{\,i}_{jk}\} + S_{jk}{}^i - S^i{}_{jk} - S^i{}_{kj}$$

by virtue of (7.4).

From (7.5), we have

$$\frac{1}{2}(E^i_{jk} + E^i_{kj}) = \{^{\,i}_{jk}\} - S^i{}_{jk} - S^i{}_{kj}$$

so that, the symmetric part of E^i_{jk} does not necessarily coincide with the Christoffel symbols $\{^{\,i}_{jk}\}$. In order that such be the case, we must have

$$S^i{}_{jk} + S^i{}_{kj} = 0$$

or

$$S_{ijk} + S_{ikj} = 0$$

Thus, the covariant torsion tensor S_{ijk}, which is by definition anti-symmetric in i and j, must be anti-symmetric in all indices. The converse being evident, we have

THEOREM 7.1. A necessary and sufficient condition that the symmetric part of E^i_{jk} coincide with the Christoffel symbols $\{^{\,i}_{jk}\}$ is that the covariant components S_{ijk} of the torsion tensor be anti-symmetric in all indices.

In the case of semi-simple group space explained in Section 1, of Chapter VI, we have

$$\Omega_{jki} = S_{jki}$$

and so since S_{jki} is anti-symmetric in all indices, we obtain

(7.6) $$\frac{1}{2}\,(E^i_{jk} + E^i_{kj}) = \{{}^{\,i}_{jk}\}$$

Now, taking a general tensor, say, $P^i{}_{jk}$, if we calculate $P^i{}_{jk|l|m} - P^i{}_{jk|m|l}$, then we find the Ricci formula:

(7.7) $$P^i{}_{jk|l|m} - P^i{}_{jk|m|l} = P^s{}_{jk}E^i{}_{slm} - P^i{}_{sk}E^s{}_{jlm} - P^i{}_{js}E^s{}_{klm} - 2P^i{}_{jk|s}S_{lm}{}^s$$

where

(7.8) $$E^i{}_{jkl} = \frac{\partial E^i_{jk}}{\partial x^l} - \frac{\partial E^i_{jl}}{\partial x^k} + E^s_{jk}E^i_{sl} - E^s_{jl}E^i_{sk}$$

is the curvature tensor of the metric connection E^i_{jk}.

Applying the Ricci formula to g_{ij}, we find

$$0 = g_{ij|k|l} - g_{ij|l|k} = -\,g_{sj}E^s{}_{ikl} - g_{is}E^s{}_{jkl}$$

and on putting

$$E_{ijkl} = g_{is}E^s{}_{jkl}$$

we obtain

(7.9) $$E_{ijkl} = -\,E_{jikl} \quad\text{and}\quad E_{ijkl} = -\,E_{ijlk}$$

It will be easily verified that the components $E^i{}_{jkl}$ of the curvature tensor satisfy, instead of the usual ones, the following Bianchi identities:

(7.10) $$E^i{}_{jkl} + E^i{}_{klj} + E^i{}_{ljk} - 2(S_{jk}{}^i{}_{|l} + S_{kl}{}^i{}_{|j} + S_{lj}{}^i{}_{|k})$$
$$+\,4(S_{jk}{}^tS_{tl}{}^i + S_{kl}{}^tS_{tj}{}^i + S_{lj}{}^tS_{tk}{}^i) = 0$$

(7.11) $$E^i{}_{jkl|m} + E^i{}_{jlm|k} + E^i{}_{jmk|l} - 2(E^i{}_{jtk}S_{lm}{}^t + E^i{}_{jtl}S_{mk}{}^t + E^i{}_{jtm}S_{kl}{}^t) = 0$$

and for a semi-simple group space with $E^i{}_{jkl} = 0$ and $\Omega_{jk}{}^i = S_{jk}{}^i$, equation (7.10) reduces to the Jacobi identity, and equation (7.11) to an identity.

Also on putting

$$(7.12)\qquad \begin{cases} E^i_{jk} = \{^{\,i}_{jk}\} + T_{jk}{}^i \\ \\ T_{jk}{}^i = S_{jk}{}^i - S^i{}_{jk} - S^i{}_{kj} \end{cases}$$

we find

$$(7.13)\qquad T_{js}{}^s = 2S_{js}{}^s$$

and

$$(7.14)\qquad E^i{}_{jkl} = R^i{}_{jkl} + T_{jk}{}^i{}_{;l} - T_{jl}{}^i{}_{;k} + T_{jk}{}^s T_{sl}{}^i - T_{jl}{}^s T_{sk}{}^i$$

where

$$R^i{}_{jkl} = \frac{\partial\{^{\,i}_{jk}\}}{\partial x^l} - \frac{\partial\{^{\,i}_{jl}\}}{\partial x^k} + \{^{\,a}_{jk}\}\{^{\,i}_{al}\} - \{^{\,a}_{jl}\}\{^{\,i}_{ak}\}$$

and the semi-colon denotes covariant differentiation with respect to the Christoffel symbols $\{^{\,i}_{jk}\}$.

As in the case of semi-simple group space, if we assume that S_{ijk} is anti-symmetric in all indices, then we have $T_{jk}{}^i = S_{jk}{}^i$ and equation (7.14) becomes

$$(7.15)\qquad E^i{}_{jkl} = R^i{}_{jkl} + S_{jk}{}^i{}_{;l} - S_{jl}{}^i{}_{;k} + S_{jk}{}^t S_{tl}{}^i - S_{jl}{}^t S_{tk}{}^i$$

and this implies

$$(7.16)\qquad E_{jk} = R_{jk} + S_{jk}{}^t{}_{;t} + S_{jl}{}^t S_{kt}{}^l$$

where

$$(7.17)\qquad E_{jk} = E^l{}_{jkl} \quad \text{and} \quad R_{jk} = R^l{}_{jkl}$$

and due to

$$g^{jk}E_{ijkl} = g^{jk}E_{jilk} = E_{il}$$

we also have

$$g^{jk}E^i{}_{jkl} = E^i{}_l \tag{7.18}$$

From (7.16), we deduce

$$\tfrac{1}{2}(E_{jk} + E_{kj}) = R_{jk} - S_j{}^{rs}S_{krs} \tag{7.19}$$

$$\tfrac{1}{2}(E_{jk} - E_{kj}) = S_{jk}{}^t{}_{;t} \tag{7.20}$$

and thus the tensor E_{jk} is not symmetric in general. But, from (7.19), we have

$$E_{jk}\xi^j\xi^k = R_{jk}\xi^j\xi^k - (S_j{}^{rs}\xi^j)(S_{krs}\xi^k) \tag{7.21}$$

and hence we have

THEOREM 7.2. In a metric manifold with anti-symmetric torsion tensor, if $E_{jk} + E_{kj} = 0$, then $R_{jk}\xi^j\xi^k$ is non-negative.

THEOREM 7.3. In a metric manifold with anti-symmetric torsion tensor, if $R_{jk}\xi^j\xi^k$ is non-positive, then $E_{jk}\xi^j\xi^k$ is also non-positive.

2. THEOREM OF HOPF-BOCHNER AND SOME APPLICATIONS

Now, in a compact manifold with positive definite metric $ds^2 = g_{jk}dx^j dx^k$ and with linear connection E^i_{jk}, we have, for a scalar function $\phi(x)$,

$$\phi_{|j} = \frac{\partial\phi}{\partial x^j}$$

$$\phi_{|j|k} = \frac{\partial^2\phi}{\partial x^j \partial x^k} - \frac{\partial\phi}{\partial x^i} E^i_{jk}$$

and consequently

$$\Delta\phi \equiv g^{jk}\phi_{|j|k} = g^{jk}\frac{\partial^2\phi}{\partial x^j \partial x^k} - g^{jk}E^i_{jk}\frac{\partial\phi}{\partial x^i}$$

and hence, applying Theorem 2.2, we obtain

THEOREM 7.4. In a compact manifold with positive definite metric, if, for a scalar $\phi(x)$, we have

$$\Delta\phi \equiv g^{jk}\phi_{|j|k} \geq 0$$

then we have

$$\Delta\phi = 0$$

As an application of this theorem, we have

THEOREM 7.5. In a compact metric manifold with torsion, if a vector ξ_i satisfies the relation

$$g^{jk}\xi_{i|j|k} = U_{ij}\xi^j + 2V_{irs}\xi^{r|s} \tag{7.22}$$

then we cannot have

$$A \equiv U_{ij}\xi^i\xi^j + 2V_{irs}\xi^i\xi^{r|s} + g_{rt}g_{su}\xi^{r|s}\xi^{t|u} \geq 0$$

unless equality holds.

More generally, if a tensor $\xi_{i_1 i_2 \cdots i_p}$ satisfies

$$g^{jk}\xi_{i_1 i_2 \cdots i_p|j|k} = U_{i_1 i_2 \cdots i_p j_1 j_2 \cdots j_p}\xi^{j_1 j_2 \cdots j_p} + 2V_{i_1 i_2 \cdots i_p r_1 r_2 \cdots r_p s}\xi^{r_1 r_2 \cdots r_p|s}$$

then we cannot have

$$\begin{aligned} A \equiv\ & U_{i_1 i_2 \cdots i_p j_1 j_2 \cdots j_p}\xi^{i_1 i_2 \cdots i_p}\xi^{j_1 j_2 \cdots j_p} \\ & + 2V_{i_1 i_2 \cdots i_p r_1 r_2 \cdots r_p s}\xi^{i_1 i_2 \cdots i_p}\xi^{r_1 r_2 \cdots r_p|s} \\ & + g_{r_1 t_1}g_{r_2 t_2}\cdots g_{r_p t_p}g_{su}\xi^{r_1 r_2 \cdots r_p|s}\xi^{t_1 t_2 \cdots t_p|u} \geq 0 \end{aligned}$$

unless equality holds.

For the proof we note that if $\phi = \xi^i\xi_i$, then we have

$$\frac{1}{2}\Delta\phi = \xi^i g^{jk}\xi_{i|j|k} + g_{rt}g_{su}\xi^{r|s}\xi^{t|u}$$

where

$$\xi^{r|s} = \xi^r{}_{|a}g^{as}$$

and thus if ξ^i satisfies (7.22), then we have

$$\frac{1}{2}\Delta\phi = A$$

and hence the conclusion by Theorem 7.4.

The extension to tensors is by analogy.

THEOREM 7.6. In a compact metric manifold with torsion, there exists no vector field which satisfies

$$g^{jk}(\xi_{i|j} - \xi_{j|i})_{|k} + \xi^j{}_{|j|i} = 0 \tag{7.23}$$

and

$$E_{ij}\xi^i\xi^j - 2S_{irs}\xi^i\xi^{r|s} + g_{rt}g_{su}\xi^{r|s}\xi^{t|u} \geq 0$$

unless equality holds.

In fact, we have the general identity:

$$g^{jk}\xi_{i|j|k} - g^{jk}(\xi_{i|j} - \xi_{j|i})_{|k} - \xi^j{}_{|j|i} = E_{ai}\xi^a - 2S_{irs}\xi^{r|s} \tag{7.24}$$

and thus if ξ^i satisfies (7.23) then it also satisfies

$$g^{jk}\xi_{i|j|k} = E_{ai}\xi^a - 2S_{irs}\xi^{r|s} \tag{7.25}$$

and now apply Theorem 7.5.

Similarly we obtain

THEOREM 7.7. In a compact metric manifold with torsion, there exists no vector field which satisfies

$$g^{jk}(\xi_{i|j} + \xi_{j|i})_{|k} - \xi^{j}{}_{|j|i} = 0 \tag{7.26}$$

and

$$E_{ij}\xi^{i}\xi^{j} - 2S_{irs}\xi^{i}\xi^{r|s} - g_{rt}g_{su}\xi^{r|s}\xi^{t|u} \leq 0$$

unless equality holds.

Next, if an anti-symmetric tensor $\xi_{i_1 i_2 \cdots i_p}$ satisfies

$$\begin{aligned} g^{jk}(\xi_{i_1 i_2 \cdots i_p|j} - \xi_{j i_2 \cdots i_p|i_1} - \cdots - \xi_{i_1 i_2 \cdots i_{p-1} j|i_p})_{|k} \\ -(\xi^{a}{}_{i_2 \cdots i_p|a|i_1} - \xi^{a}{}_{i_1 i_3 \cdots i_p|a|i_2} - \cdots - \xi^{a}{}_{i_2 \cdots i_{p-1} i_1|a|i_p}) = 0 \end{aligned} \tag{7.27}$$

then, for $\phi = \xi^{i_1 i_2 \cdots i_p}\xi_{i_1 i_2 \cdots i_p}$, we have

$$\begin{aligned} \Delta\phi = K^{(p)}_{ijkl}\xi^{iji_3 \cdots i_p}\xi^{kl}{}_{i_3 \cdots i_p} - 2S^{(p)}_{ijrst}\xi^{iji_3 \cdots i_p}\xi^{rs}{}_{i_3 \cdots i_p}{}^{|t} \\ + G_{rstuvw}\xi^{rsi_3 \cdots i_p|t}\xi^{uv}{}_{i_3 \cdots i_p}{}^{|w} \end{aligned} \tag{7.28}$$

where

$$\begin{aligned} K^{(p)}_{ijkl} = \frac{p}{2}(E_{ik}g_{lj} - E_{jk}g_{li} - E_{il}g_{kj} + E_{jl}g_{ki}) - \frac{p(p-1)}{2}(E_{iklj} - E_{jkli} \\ - E_{ilkj} + E_{jlki}), \end{aligned}$$

$$S^{(p)}_{ijrst} = \frac{p}{2}(S_{irt}g_{js} - S_{jrt}g_{is} - S_{ist}g_{jr} + S_{jst}g_{ir}),$$

$$G_{rstuvw} = (g_{ru}g_{sv} - g_{rv}g_{su})g_{tw},$$

and if, on the other hand, the tensor satisfies

$$\begin{aligned} g^{jk}(p\xi_{i_1 i_2 \cdots i_p|j} + \xi_{j i_2 \cdots i_p|i_1} + \cdots + \xi_{i_1 i_2 \cdots i_{p-1} j|i_p})_{|k} \\ -(\xi^{a}{}_{i_2 \cdots i_p|a|i_1} - \xi^{a}{}_{i_1 i_3 \cdots i_p|a|i_2} - \cdots - \xi^{a}{}_{i_2 \cdots i_{p-1} i_1|a|i_p}) = 0 \end{aligned} \tag{7.29}$$

then we have

$$(7.30)\quad \Delta\phi = -\frac{1}{p}\Big[K^{(p)}_{ijkl}\xi^{iji_3\ldots i_p}\xi^{kl}{}_{i_3\ldots i_p} - 2S^{(p)}_{ijrst}\xi^{iji_3\ldots i_p}\xi^{rs}{}_{i_3\ldots i_p}{}^{|t}$$
$$- pG_{rstuvw}\xi^{rsi_3\ldots i_p|t}\xi^{uv}{}_{i_3\ldots i_p}{}^{|w}\Big]$$

THEOREM 7.8. In a compact metric manifold with torsion, there is no anti-symmetric tensor which satisfies (7.27) and for which

$$K^{(p)}_{ijkl}\xi^{iji_3\ldots i_p}\xi^{kl}{}_{i_3\ldots i_p} - 2S^{(p)}_{ijrst}\xi^{iji_3\ldots i_p}\xi^{rs}{}_{i_3\ldots i_p}{}^{|t}$$
$$+ G_{rstuvw}\xi^{rsi_3\ldots i_p|t}\xi^{uv}{}_{i_3\ldots i_p}{}^{|w} \geq 0$$

unless equality holds.

Also, if, it satisfies (7.29), then we cannot have

$$K^{(p)}_{ijkl}\xi^{iji_3\ldots i_p}\xi^{kl}{}_{i_3\ldots i_p} - 2S^{(p)}_{ijrst}\xi^{iji_3\ldots i_p}\xi^{rs}{}_{i_3\ldots i_p}{}^{|t}$$
$$- pG_{rstuvw}\xi^{rsi_3\ldots i_p|t}\xi^{uv}{}_{i_3\ldots i_p}{}^{|w} \leq 0$$

unless equality holds.

3. PSEUDO-HARMONIC VECTORS AND TENSORS

We shall call a vector pseudo-harmonic, if

$$(7.31)\qquad \xi_{i|j} = \xi_{j|i} \quad \text{and} \quad \xi^k{}_{|k} = 0$$

Such a vector satisfies evidently (7.23) and consequently (7.25), and for $\phi = \xi^i\xi_i$, we have

$$(7.32)\qquad \Delta\phi = (E_{jk} + E_{kj})\xi^j\xi^k - 2(S_{irs} + S_{isr})\xi^i\xi^{r|s}$$
$$+ (g_{rt}g_{su} + g_{ru}g_{st})\xi^{r|s}\xi^{t|u}$$

thus we obtain

THEOREM 7.9. In a compact metric manifold with torsion, if the symmetric matrix

$$M = \begin{bmatrix} E_{jk} + E_{kj} & -(S_{irs} + S_{isr}) \\ -(S_{irs} + S_{isr}) & g_{rt}g_{su} + g_{ru}g_{st} \end{bmatrix}$$

defines a non-negative quadratic form in the variables ξ^i and $\xi^{rs} = \xi^{sr}$, then every pseudo-harmonic vector ξ^i must satisfy

$$(E_{jk} + E_{kj})\xi^j\xi^k - 2(S_{irs} + S_{isr})\xi^i\xi^{r|s} + (g_{rt}g_{su} + g_{ru}g_{st})\xi^{r|s}\xi^{t|u} = 0$$

If the matrix M defines a positive definite form, then there exists no pseudo-harmonic vector other than zero.

Now, if we have

$$E_{jk} + E_{kj} = 0 \qquad \text{and} \qquad S_{irs} + S_{isr} = 0$$

then, for a pseudo-harmonic vector ξ^i, we have

$$\Delta\phi = 2\xi^{j|k}\xi_{j|k} = 0$$

from which

$$\xi_{j|k} \equiv \frac{\partial \xi_j}{\partial x^k} - \xi_i E^i_{jk} = 0$$

follows. Thus, there exists in this case at most n linearly independent (with constant coefficients) pseudo-harmonic vectors. Moreover, if such a pseudo-harmonic vector exists, it must satisfy

$$\xi_{j|k} = \xi_{j;k} - \xi_i S_{jk}{}^i = 0$$

from which

$$\xi_{j;k} + \xi_{k;j} = 0$$

The last equation shows that ξ^i is an ordinary Killing vector, and thus we have

THEOREM 7.10. In a compact metric manifold with torsion satisfying $E_{jk} + E_{kj} = 0$ and $S_{irs} + S_{isr} = 0$, a pseudo-harmonic vector must have vanishing covariant derivative with respect to the connection of the manifold, and consequently the number of linearly independent (with constant coefficients) pseudo-harmonic vectors is at most n. Moreover, if such a pseudo-harmonic vector exists, it is then an ordinary Killing vector.

Now, if $E_{jk} + E_{kj} = 0$ and $S_{irs} + S_{isr} = 0$, then by Theorem 7.2, $R_{jk}\xi^j\xi^k$ is non-negative. Hence by Theorem 2.9, an ordinary harmonic vector must have vanishing covariant derivative with respect to the Christoffel symbols and satisfy

$$R_{jk}\xi^j\xi^k = S_{jrs}S_k{}^{rs}\xi^j\xi^k = 0$$

Thus, if the rank of the matrix $\| S_{jrs}S_k{}^{rs} \|$ is n, we can conclude that there exists no ordinary harmonic vector. Thus, we have

THEOREM 7.11. In a compact metric manifold with torsion satisfying $E_{jk} + E_{kj} = 0$ and $S_{irs} + S_{isr} = 0$, the ordinary harmonic vector must have vanishing covariant derivative with respect to the Christoffel symbols. Moreover if the rank of the matrix $\| S_{jrs}S_k{}^{rs} \|$ is n, then there exists no ordinary harmonic vector.

A compact semi-simple group space falls under Theorem 7.10 and 7.11. On the other hand, in such a space, a pseudo-harmonic vector ξ_i can be written as

$$\xi_j = f_a(x)h^a_j$$

and by Theorem 7.10, it must have a vanishing covariant derivative with respect to the connection of the manifold; consequently, the covariant derivatives of h^a_j being zero, $f_a(x)$ must be constants. Thus we have

THEOREM 7.12. In a compact semi-simple group space, there exist n linearly independent pseudo-harmonic vectors, and any pseudo-harmonic vector is a linear combination with constant coefficients of these vectors.

Moreover, from

$$0 = h^a_{j|k} = h^a_{j;k} - h^a_i S_{jk}{}^i$$

we have

$$h^a_{j;k} + h^a_{k;j} = 0$$

and thus h^a_j are all ordinary Killing vectors and the manifold admits a simply transitive motions as well known.

Now, we shall call an anti-symmetric tensor

$$\xi_{i_1 i_2 \cdots i_p}$$

pseudo-harmonic if it satisfies the conditions:

(7.33) $$\xi_{[i_1 i_2 \cdots i_p | r]} = 0$$

or explicitly

(7.34) $$\xi_{i_1 i_2 \cdots i_p | r} = \xi_{r i_2 \cdots i_p | i_1} + \xi_{i_1 r i_3 \cdots i_p | i_2} + \cdots + \xi_{i_1 i_2 \cdots i_{p-1} r | i_p}$$

and

(7.35) $$g^{rs} \xi_{r i_2 \cdots i_p | s} = 0$$

where the brackets [ ] denote the anti-symmetric part.

Such an anti-symmetric tensor evidently satisfies (7.27) and consequently (7.28), and we may state as follows:

THEOREM 7.13. The first half of Theorem 7.8 applies in particular to pseudo-harmonic tensors.

4. PSEUDO-KILLING VECTORS AND TENSORS

We shall call a vector ξ^i pseudo-Killing vector, if it satisfies the condition

$$\xi_{i|j} + \xi_{j|i} = 0 \quad \text{and automatically} \quad \xi^i{}_{|i} = 0 \tag{7.36}$$

Such a vector evidently satisfies

$$g^{jk}(\xi_{i|j} + \xi_{j|i})_{|k} - \xi^j{}_{|j|i} = 0$$

and consequently

$$g^{jk}\xi_{i|j|k} = -E_{ai}\xi^a + 2S_{irs}\xi^{r|s}$$

and the analogue to Theorem 7.9 is as follows:

THEOREM 7.14. *In a compact metric manifold with torsion if the matrix*

$$M' = \begin{bmatrix} E_{jk} + E_{kj} & -(S_{irs} - S_{isr}) \\ -(S_{irs} - S_{isr}) & -(g_{rt}g_{su} - g_{ru}g_{st}) \end{bmatrix}$$

defines a non-positive quadratic form in the variables ξ^i and $\xi^{rs} = -\xi^{sr}$, then every pseudo-Killing vector ξ^i must satisfy

$$(E_{jk} + E_{kj})\xi^j\xi^k - 2(S_{irs} - S_{isr})\xi^i\xi^{r|s} - (g_{rt}g_{su} - g_{ru}g_{st})\xi^{r|s}\xi^{t|u} = 0$$

If the matrix M' defines a negative definite form, then there exists no pseudo-Killing vector other than zero vector.

We shall call an anti-symmetric tensor field

$$\xi_{i_1 i_2 \ldots i_p}$$

pseudo-Killing tensor if it satisfies the conditions:

$$\xi_{i_1 i_2 \dots i_p | r} = \xi_{[i_1 i_2 \dots i_p | r]} \tag{7.37}$$

or explicitly

$$\xi_{i_1 i_2 \dots i_p | r} = -\frac{1}{p}\left(\xi_{r i_2 \dots i_p | i_1} + \xi_{i_1 r i_3 \dots i_p | i_2} + \dots + \xi_{i_1 i_2 \dots i_{p-1} r | i_p}\right) \tag{7.38}$$

and automatically

$$g^{rs}\xi_{r i_2 \dots i_p | s} = 0 \tag{7.39}$$

Such an anti-symmetric tensor evidently satisfies

$$g^{jk}(p\xi_{i_1 i_2 \dots i_p | j} + \xi_{j i_2 \dots i_p | i_1} + \dots + \xi_{i_1 i_2 \dots i_{p-1} j | i_p})_{|k}$$
$$- (\xi^{a}{}_{i_2 \dots i_p | a | i_1} - \xi^{a}{}_{i_1 i_3 \dots i_p | a | i_2} - \dots - \xi^{a}{}_{i_2 \dots i_{p-1} i_1 | a | i_p}) = 0$$

and consequently, for

$$\phi = \xi^{i_1 i_2 \dots i_p}\xi_{i_1 i_2 \dots i_p}$$

we get

$$\Delta\phi = -\frac{1}{p}\left[K^{(p)}_{ijkl}\xi^{iji_3 \dots i_p}\xi^{kl}{}_{i_3 \dots i_p} - 2S^{(p)}_{ijrst}\xi^{iji_3 \dots i_p}\xi^{rs}{}_{i_3 \dots i_p}{}^{|t} - pG_{rstuvw}\xi^{rsi_3 \dots i_p | t}\xi^{uv}{}_{i_3 \dots i_p}{}^{|w}\right]$$

and thus we have

THEOREM 7.15. The second half of Theorem 7.8 applies in particular to pseudo-Killing tensors.

5. INTEGRAL FORMULAS

In this section, we shall consider a compact orientable metric manifold with torsion, and suppose that the torsion tensor $S_{jk}{}^{i}$ satisfies the condition

$$S_{ji}{}^{i} = 0 \tag{7.40}$$

and this condition is satisfied automatically if the covariant torsion tensor S_{jki} is anti-symmetric in all indices.

First, for any vector v^i, we have

$$v^i{}_{|i} = v^i{}_{;i} + v^j T_{ji}{}^{i} = v^i{}_{;i} + 2v^j S_{ji}{}^{i}$$

from which

$$v^i{}_{|i} = \frac{1}{\sqrt{g}} \frac{\partial \sqrt{g}\, v^i}{\partial x^i} \tag{7.41}$$

by virtue of the assumption (7.40), where g is the determinant formed with g_{jk}. Thus, for any vector field $v^i(x)$, we have

$$\int v^i{}_{|i} dv = 0 \tag{7.42}$$

the integral being taken over the whole manifold, where dv is the volume element.

Applying first (7.42) to the vector $\xi^i{}_{|j}\xi^j$, we find

$$\int (\xi^i{}_{|j}\xi^j)_{|i} dv = \int (\xi^i{}_{|j|i}\xi^j + \xi^i{}_{|j}\xi^j{}_{|i}) dv = 0$$

or

$$\int (\xi^i{}_{|i|j}\xi^j + E_{jk}\xi^j\xi^k - 2\xi^i{}_{|s}S_{ji}{}^{s}\xi^j + \xi^i{}_{|j}\xi^j{}_{|i}) dv = 0 \tag{7.43}$$

by virtue of Ricci identity:

$$\xi^i{}_{|j|k} - \xi^i{}_{|k|j} = \xi^s E^i{}_{sjk} - 2\xi^i{}_{|s}S_{jk}{}^{s}$$

Applying it next to the vector $\xi^i{}_{|i}\xi^j$, we have

(7.44) $$\int(\xi^i{}_{|i}\xi^j)_{|j}dv = \int(\xi^i{}_{|i|j}\xi^j + \xi^i{}_{|i}\xi^j{}_{|j})dv = 0$$

and hence

(7.45) $$\int(E_{jk}\xi^j\xi^k - 2\xi^i{}_{|s}S_{ji}{}^s\xi^j + \xi^i{}_{|j}\xi^j{}_{|i} - \xi^i{}_{|i}\xi^j{}_{|j})dv = 0$$

If the vector ξ_i is pseudo-harmonic, then equation (7.45) becomes

$$\int[(E_{jk} + E_{kj})\xi^j\xi^k - 2(S_{irs} + S_{isr})\xi^i\xi^{r|s} + (g_{rt}g_{su} + g_{ru}g_{st})\xi^{r|s}\xi^{t|u}]dv = 0$$

and this gives another proof of Theorem 7.9 for a compact orientable metric manifold with torsion satisfying $S_{ji}{}^i = 0$.

If the vector ξ_i is pseudo-Killing, then equation (7.45) becomes

$$\int[(E_{jk} + E_{kj})\xi^j\xi^k - 2(S_{irs} - S_{isr})\xi^i\xi^{r|s} - (g_{rt}g_{su} - g_{ru}g_{st})\xi^{r|s}\xi^{t|u}]dv = 0$$

and this gives another proof of Theorem 7.14 for a compact orientable metric manifold with torsion satisfying $S_{ji}{}^i = 0$.

The generalization of formula (7.45) to the case of anti-symmetric tensor is

(7.46) $$\int[E_{ij}\xi^{ii_2\cdots i_p}\xi^j{}_{i_2\cdots i_p} + (p - 1)E_{ljki}\xi^{iji_3\cdots i_p}\xi^{kl}{}_{i_3\cdots i_p}$$
$$- 2S_{irs}\xi^{ii_2\cdots i_p}\xi^r{}_{i_2\cdots i_p}{}^{|s} + \frac{1}{p}\xi^{i_1i_2\cdots i_p|j}\xi_{i_1i_2\cdots i_p|j}$$
$$- \frac{p+1}{p}\xi^{[i_1i_2\cdots i_p|j]}\xi_{[i_1i_2\cdots i_p|j]} - \xi^{ii_2\cdots i_p}{}_{|i}\xi^j{}_{i_2\cdots i_p|j}]dv = 0$$

and Theorems 7.13 and 7.15 can be again reproved for $S_{ji}{}^i = 0$.

6. NECESSARY AND SUFFICIENT CONDITION THAT A TENSOR BE A PSEUDO-HARMONIC OR PSEUDO-KILLING TENSOR

Under the same assumption as in Section 5, we have

$$\int g^{jk}(\xi^i \xi_i)_{|j|k} dv = 0$$

for any vector field ξ^i , or

$$(7.47) \qquad \int (\xi^{i|j} \xi_{i|j} + \xi^i g^{jk} \xi_{i|j|k}) dv = 0$$

and hence

$$(7.48) \qquad \int [\xi^i (g^{jk} \xi_{i|j|k} - \xi_s E^s{}_i + 2S_{irs} \xi^{r|s})$$

$$+ \frac{1}{2} (\xi^{i|j} - \xi^{j|i})(\xi_{i|j} - \xi_{j|i}) + \xi^i{}_{|i} \xi^j{}_{|j}] dv = 0$$

Thus, if the vector ξ^i satisfies

$$(7.49) \qquad g^{jk} \xi_{i|j|k} - \xi_s E^s{}_i + 2S_{irs} \xi^{r|s} = 0$$

then we must have

$$\xi_{i|j} - \xi_{j|i} = 0 \quad \text{and} \quad \xi^i{}_{|i} = 0$$

that is to say, the ξ_i must be pseudo-harmonic. Conversely, if ξ_i is pseudo-harmonic, then we have (7.49). Thus we have

THEOREM 7.16. In a compact orientable metric manifold with torsion satisfying $S_{ji}{}^i = 0$, a necessary and sufficient condition that a vector be pseudo-harmonic vector is that the ξ_i satisfy (7.49).

Similarly we find

$$(7.50) \qquad \int [\xi^i (g^{jk} \xi_{i|j|k} + \xi_s E^s{}_i - 2S_{irs} \xi^{r|s})$$

$$+ \frac{1}{2} (\xi^{i|j} + \xi^{j|i})(\xi_{i|j} + \xi_{j|i}) - \xi^i{}_{|i} \xi^j{}_{|j}] dv = 0$$

and thus, if the vector ξ_i satisfies

$$(7.51) \qquad g^{jk} \xi_{i|j|k} + \xi_s E^s{}_i - 2S_{irs} \xi^{r|s} = 0$$

and

$$\xi^i{}_{|i} = 0 \tag{7.52}$$

then we must have

$$\xi_{i|j} + \xi_{j|i} = 0 \tag{7.53}$$

THEOREM 7.17. In a compact orientable metric manifold with torsion satisfying $S_{ji}{}^i = 0$, a necessary and sufficient condition that a vector ξ_i be pseudo-Killing vector is that ξ_i satisfy (7.51) and (7.52).

As generalizations of Theorems 7.16 and 7.17, one can prove

THEOREM 7.18. In a compact orientable metric manifold whose torsion tensor $S_{jk}{}^i$ satisfies $S_{ji}{}^i = 0$, an anti-symmetric tensor

$$\xi_{i_1 i_2 \cdots i_p}$$

is pseudo-harmonic if and only if

$$g^{rs}\xi_{i_1 i_2 \cdots i_p|r|s} - \sum_{s=1}^{p} \xi_{i_1 \cdots i_{s-1} a i_{s+1} \cdots i_p} E^a{}_{i_s}$$
$$+ \sum_{s<t}^{1\cdots p} \xi_{i_1 \cdots i_{s-1} a i_{s+1} \cdots i_{t-1} b i_{t+1} \cdots i_p} (E^a{}_{i_s i_t}{}^b - E^a{}_{i_t i_s}{}^b)$$
$$+ 2\sum_{s=1}^{p} \xi_{i_1 \cdots i_{s-1} r i_{s+1} \cdots i_p|t} S_{i_s}{}^{rt} = 0$$

holds, and is a pseudo-Killing tensor if and only if

$$g^{rs}\xi_{i_1 i_2 \cdots i_p|r|s} + \frac{1}{p}\sum_{s=1}^{p} \xi_{i_1 \cdots i_{s-1} a i_{s+1} \cdots i_p} E^a{}_{i_s}$$
$$- \frac{1}{p}\sum_{s<t}^{1\cdots p} \xi_{i_1 \cdots i_{s-1} a i_{s+1} \cdots i_{t-1} b i_{t+1} \cdots i_p} (E^a{}_{i_s i_t}{}^b - E^a{}_{i_t i_s}{}^b)$$
$$- \frac{2}{p}\sum_{s=1}^{p} \xi_{i_1 \cdots i_{s-1} r i_{s+1} \cdots i_p|t} S_{i_s}{}^{rt} = 0$$

and

$$\xi^{a}{}_{i_2 \ldots i_p | a} = 0$$

holds.

7. A GENERALIZATION

Theorem 7.4 can obviously be generalized as follows:

THEOREM 7.17. On a compact metric manifold with torsion, if, for a scalar ϕ and a vector A^s, we have

$$\Delta\phi + 2A^s \phi_{|s} \equiv g^{jk}\phi_{|j|k} + 2A^s\phi_{|s} \geq 0$$

then we have

$$\Delta\phi + 2A^s\phi_{|s} = 0$$

By introducing the value from $\Delta\phi$ we may now conclude as follows:

THEOREM 7.18. In a compact metric manifold with torsion, for a pseudo-harmonic vector ξ^i, we cannot have the inequality

$$(E_{jk} + E_{kj})\xi^j\xi^k - 2(S_{irs} + S_{isr} - g_{ir}A_s - g_{is}A_r)\xi^i\xi^{r|s}$$
$$+ (g_{rt}g_{su} + g_{ru}g_{st})\xi^{r|s}\xi^{t|u} \geq 0$$

for any vector field A_s whatsoever, unless equality holds.

Also, for a pseudo-Killing vector ξ^i, we cannot have the inequality

$$(E_{jk} + E_{kj})\xi^j\xi^k - 2(S_{irs} - S_{isr} + g_{ir}A_s - g_{is}A_r)\xi^i\xi^{r|s}$$
$$- (g_{rt}g_{su} - g_{ru}g_{st})\xi^{r|s}\xi^{t|u} \leq 0$$

for any vector field A_s whatsoever, unless equality holds.

In the particular case where the torsion tensor satisfies an equation of the form

$$S_{irs} - S_{isr} + g_{ir}A_s - g_{is}A_r = 0 \tag{7.54}$$

we have

THEOREM 7.19. In a compact metric manifold with torsion tensor $S_{jk}{}^{i}$ satisfying an equation of the form (7.54), if $E_{jk}\xi^j\xi^k$ is non-positive, then every pseudo-Killing vector ξ_i must have vanishing covariant derivative with respect to the affine connection of the manifold.

If $E_{jk}\xi^j\xi^k$ is negative definite, then there exists no pseudo-Killing vector other than the zero vector.

It is to be noted that if $S_{jk}{}^{i}$ has the form $S_{jk}{}^{i} = \delta^i_j\phi_k - \delta^i_k\phi_j$, then (7.54) is satisfied.

Such an unspecified vector field A_s can also be introduced in a suitable fashion in the formulas in Theorems 7.13 and 7.15 without altering the wording of the theorems otherwise.

CHAPTER VIII

KAEHLER MANIFOLD

1. KAEHLER MANIFOLD

Consider a real 2n-dimensional manifold V_{2n} of class C^r with a given covering by neighborhoods each endowed with a coordinate system.

If we denote the coordinates of a point P in two different coordinate neighborhoods by (x^i) and (x'^i) respectively, then there exist relations of the form

$$x'^i = x'^i(x^a) \tag{8.1}$$

where the functions $x'^i(x^a)$ are functions of class C^r with non-vanishing Jacobian, and the Latin indices take the values $1, 2, \ldots, n$; $\bar{1}, \bar{2}, \ldots, \bar{n}$.

Now, if we put

$$\begin{cases} z^\alpha = x^\alpha + ix^{\bar{\alpha}} \\ \bar{z}^\alpha = x^\alpha - ix^{\bar{\alpha}} \end{cases} \tag{8.2}$$

where Greek indices take the values $1, 2, \ldots, n$, then we have a one-to-one correspondence

$$(z^\alpha, \bar{z}^\alpha) \rightleftarrows (x^i)$$

and $(z^\alpha, \bar{z}^\alpha)$ may be considered to be coordinates of a point in our real 2n-dimensional manifold V_{2n}, and equations (8.1) may be always written as relations

$$z'^\alpha = z'^\alpha(z, \bar{z}) \qquad \bar{z}'^\alpha = \bar{z}'^\alpha(z, \bar{z}) \tag{8.3}$$

formally.

Now, if

(I) We can cover the manifold entirely by a system of coordinate neighborhoods endowed with complex coordinates $(z^\alpha, \bar{z}^\alpha)$,

and if

(II) U_1 and U_2 being two complex coordinate neighborhoods of the manifold, if a point P belongs to $U_1 \cap U_2$, then the complex coordinates z'^α of the point P in one of these complex coordinate neighborhoods are complex analytic functions with non-vanishing Jacobian of the complex coordinates z^α of the same point,

then we say that the manifold has a complex analytic structure, and we call the manifold a complex analytic (or simply analytic) manifold of real dimension $2n$ and of complex dimension n.

In this case, equations (8.3) take the form

$$z'^\alpha = \psi^\alpha(z) \qquad \bar{z}'^\alpha = \bar{\psi}^\alpha(\bar{z}) \tag{8.4}$$

where $\bar{\psi}^\alpha$ denotes complex conjugate of the function $\psi^\alpha(z)$. Also if we put

$$\bar{z}^\alpha = z^{\bar{\alpha}}$$

and assume that barred Greek indices take the values $\bar{1}, \bar{2}, \ldots, \bar{n}$, then for $(z^\alpha, \bar{z}^\alpha)$ we can write z^i, $i = 1, 2, \ldots, n; \bar{1}, \bar{2}, \ldots, \bar{n}$, and the transformation (8.4) as

$$z'^i = f^i(z^a) \tag{8.5}$$

The Jacobian of (8.4) is then, as easily seen,

$$\left|\frac{\partial z'^i}{\partial z^j}\right| = \text{real} > 0$$

and thus the manifold is always orientable.

We shall denote a complex analytic manifold of complex dimension n by C_n.

In C_n, vectors, tensors, affine connections, etc., are defined with respect to the coordinate transformations (8.5) having the special form (8.4), in just the same way as in real case.

For example, for a contravariant vector ξ^i the law of transformation of its components:

$$\xi'^i = \frac{\partial z'^i}{\partial z^r}\xi^r$$

separates into

$$\xi'^\alpha = \frac{\partial z'^\alpha}{\partial z^\lambda}\xi^\lambda \quad \text{and} \quad \xi'^{\bar\alpha} = \frac{\partial \bar z'^\alpha}{\partial \bar z^\lambda}\xi^{\bar\lambda} \tag{8.6}$$

by virtue of the special form (8.4) of the transformation (8.5); and thus, the 2n components of ξ^i separate completely into the blocks ξ^α and $\xi^{\bar\alpha}$.

Equation (8.6) shows that if ξ^i are components of a contravariant vector, then $(\xi^\alpha, 0)$ and $(0, \xi^{\bar\alpha})$ are also components of contravariant vectors. Moreover, taking the complex conjugate of equation (8.6), we get

$$\overline{\xi'^{\bar\alpha}} = \frac{\partial z'^\alpha}{\partial z^\lambda}\overline{\xi^{\bar\lambda}} \quad \text{and} \quad \overline{\xi'^{\alpha}} = \frac{\partial \bar z'^\alpha}{\partial \bar z^\lambda}\overline{\xi^{\lambda}}$$

which show that, if ξ^i are components of a contravariant vector, then $(\overline{\xi^{\bar\alpha}}, \overline{\xi^{\alpha}})$ are also components of a contravariant vector.

For tensors, the same arguments apply. For example, if we are given a tensor $T^i{}_{jk}$, then

$$\delta^i_\alpha\delta^\beta_j\delta^\gamma_k T^\alpha{}_{\beta\gamma}, \qquad \delta^i_{\bar\alpha}\delta^\beta_j\delta^\gamma_k T^{\bar\alpha}{}_{\beta\gamma}, \qquad \delta^i_\alpha\delta^{\bar\beta}_j\delta^\gamma_k T^\alpha{}_{\bar\beta\gamma}, \qquad \ldots$$

are all components of tensors of the same kind as the original one.

If $\xi_{i_1 i_2 \ldots i_p}$ are components of an anti-symmetric tensor, then

$$p!\,\delta^{\alpha_1}_{[i_1}\delta^{\alpha_2}_{i_2}\cdots\delta^{\alpha_{p-h}}_{i_{p-h}}\delta^{\bar\alpha_{p-h+1}}_{i_{p-h+1}}\cdots\delta^{\bar\alpha_p}_{i_p]}\,\xi_{\alpha_1\alpha_2\ldots\alpha_{p-h}\bar\alpha_{p-h+1}\ldots\bar\alpha_p} \tag{8.7}$$

are also components of an anti-symmetric tensor of the same kind as the original one, and it is a tensor whose non zero components contain just p - h unbarred and h barred indices. We call such a tensor a pure tensor of type h .

Any anti-symmetric tensor can be expressed as a sum of pure tensors of type $0, 1, \ldots, p-1$ and p .

Moreover, if $T^i{}_{jk}$ are components of a tensor, then

$$S^i{}_{jk} = \overline{T^{\bar{i}}{}_{\bar{j}\bar{k}}} \tag{8.8}$$

are also components of a tensor of the same kind as the original one, where

$$\bar{i} = \begin{cases} \bar{\alpha} & \text{if } i = \alpha \\ \alpha & \text{if } i = \bar{\alpha} \end{cases}$$

We denote (8.8) by

$$S^i{}_{jk} = C(T^i{}_{jk}) \tag{8.9}$$

and we call $S^i{}_{jk}$ the adjoint of $T^i{}_{jk}$, and we call any quantity self-adjoint, if

$$T = C(T) \tag{8.10}$$

that is to say, if barring and unbarring all indices simultaneously changes the value of a component into its complex conjugate value.

For instance, a contravariant vector ξ^i is self-adjoint if it satisfies

$$\xi^{\bar{\alpha}} = \overline{\xi^{\alpha}}$$

a covariant vector ξ_i is self-adjoint if it satisfies

$$\xi_{\bar{\alpha}} = \overline{\xi_{\alpha}}$$

and a symmetric covariant tensor g_{ij} is self-adjoint if it satisfies

$$g_{\bar{\alpha}\bar{\beta}} = g_{\bar{\beta}\bar{\alpha}} = \overline{g_{\alpha\beta}} = \overline{g_{\beta\alpha}} \;,$$

$$g_{\alpha\bar{\beta}} = g_{\bar{\beta}\alpha} = \overline{g_{\bar{\alpha}\beta}} = \overline{g_{\beta\bar{\alpha}}}$$

From the complete separation of the components of tensors, it is clear that the self-adjointness is preserved by a coordinate transformation of the form (8.4). Moreover, if the covariant tensor g_{ij} with $|g_{ij}| \neq 0$ is self-adjoint, then so is the contravariant tensor g^{ij} defined by $g^{ij}g_{jk} = \delta^i_k$, and the Christoffel symbols

$$\Gamma^i_{jk} = \frac{1}{2} g^{ir} \left(\frac{\partial g_{rj}}{\partial x^k} + \frac{\partial g_{rk}}{\partial x^j} - \frac{\partial g_{jk}}{\partial x^r} \right) \tag{8.11}$$

the Riemann-Christoffel curvature tensor

$$R^i{}_{jkl} = \frac{\partial \Gamma^i_{jk}}{\partial x^l} - \frac{\partial \Gamma^i_{jl}}{\partial x^k} + \Gamma^r_{jk}\Gamma^i_{rl} - \Gamma^r_{jl}\Gamma^i_{rk} \tag{8.12}$$

$$R_{ijkl} = g_{ir}R^r{}_{jkl} \tag{8.13}$$

the Ricci tensor

$$R_{jk} = g^{il}R_{ijkl} \tag{8.14}$$

and the scalar curvature

$$R = g^{jk}R_{jk} \tag{8.15}$$

are all self-adjoint. Here, and always, a scalar is self-adjoint, if it is real valued.

Thus, if we denote the covariant differentiation with respect to Γ^i_{jk} by a semi-colon:

$$\xi^i{}_{;k} = \frac{\partial \xi^i}{\partial x^k} + \xi^j \Gamma^i_{jk}$$

then, we can see that the self-adjointness is preserved by a covariant differentiation.

Assume now that, in our complex analytic manifold, there is given a positive definite quadratic differential form

$$ds^2 = g_{jk}dz^j dz^k \tag{8.16}$$

where the symmetric tensor g_{jk} is self-adjoint and satisfies

$$g_{\alpha\beta} = g_{\bar{\alpha}\bar{\beta}} = 0 \tag{8.17}$$

From the complete separation of the components g_{jk} into four blocks $g_{\alpha\beta}$, $g_{\alpha\bar\beta}$, $g_{\bar\alpha\beta}$, $g_{\bar\alpha\bar\beta}$, it is evident that conditions (8.17) are preserved by any coordinate transformation of the form (8.4). Also, by virtue of condition (8.17), the metric form (8.16) can be written in the form

$$ds^2 = 2g_{\alpha\bar\beta}dz^\alpha d\bar z^\beta \tag{8.18}$$

where

$$g_{\alpha\bar\beta} = g_{\bar\beta\alpha} = \overline{g_{\bar\alpha\beta}} = \overline{g_{\beta\bar\alpha}} \tag{8.19}$$

and a metric (8.18) satisfying (8.19) is called a Hermitian metric.

Taking account of

$$g^{\alpha\beta} = g^{\bar\alpha\bar\beta} = 0 ,$$

$$g^{\alpha\bar\beta} = g^{\bar\beta\alpha} = \overline{g^{\bar\alpha\beta}} = \overline{g^{\beta\bar\alpha}}$$

we obtain for the Christoffel symbols Γ^i_{jk} the relations

$$\Gamma^\alpha_{\beta\gamma} = \frac{1}{2} g^{\alpha\bar\epsilon}\left(\frac{\partial g_{\bar\epsilon\beta}}{\partial z^\gamma} + \frac{\partial g_{\bar\epsilon\gamma}}{\partial z^\beta}\right)$$

$$\Gamma^\alpha_{\beta\bar\gamma} = \frac{1}{2} g^{\alpha\bar\epsilon}\left(\frac{\partial g_{\beta\bar\epsilon}}{\partial \bar z^\gamma} - \frac{\partial g_{\beta\bar\gamma}}{\partial \bar z^\epsilon}\right)$$

$$\Gamma^\alpha_{\bar\beta\bar\gamma} = 0$$

and the values of other components are given by symmetry and self-adjointness.

From the law of transformation

$$\Gamma'^i_{jk} = \frac{\partial z'^i}{\partial z^p}\left(\frac{\partial z^q}{\partial z'^j}\frac{\partial z^r}{\partial z'^k}\Gamma^p_{qr} + \frac{\partial^2 z^p}{\partial z'^j \partial z'^k}\right)$$

we get

$$\Gamma'^\alpha_{\beta\bar\gamma} = \frac{\partial z'^\alpha}{\partial z^\lambda}\frac{\partial z^\mu}{\partial z'^\beta}\frac{\partial \bar z^\nu}{\partial \bar z'^\gamma}\Gamma^\lambda_{\mu\bar\nu}$$

and thus the condition

$$\Gamma^{\alpha}_{\beta\bar{\gamma}} = 0 \tag{8.20}$$

is invariant under a coordinate transformation of the form (8.4). This is equivalent to

$$\frac{\partial g_{\alpha\bar{\beta}}}{\partial \bar{z}^{\gamma}} = \frac{\partial g_{\alpha\bar{\gamma}}}{\partial \bar{z}^{\beta}} \tag{8.21}$$

or to

$$\frac{\partial g_{\bar{\alpha}\beta}}{\partial z^{\gamma}} = \frac{\partial g_{\bar{\alpha}\gamma}}{\partial z^{\beta}} \tag{8.22}$$

or, further to

$$g_{\alpha\bar{\beta}} = \frac{\partial^2 \Phi}{\partial z^{\alpha}\partial \bar{z}^{\beta}} \tag{8.23}$$

The self-adjointness of $g_{\alpha\bar{\beta}}$ demands that Φ be a real valued function.

The condition (8.20) or (8.21) or (8.22) or (8.23) is called Kaehler's condition, and a metric satisfying (8.19) and (8.21) will be called a Kaehler metric.

Thus, in a Kaehler metric, we have

$$\Gamma^{\alpha}_{\beta\gamma} = g^{\alpha\bar{\epsilon}} \frac{\partial g_{\bar{\epsilon}\beta}}{\partial z^{\gamma}} , \qquad \Gamma^{\bar{\alpha}}_{\bar{\beta}\bar{\gamma}} = g^{\bar{\alpha}\epsilon} \frac{\partial g_{\epsilon\bar{\beta}}}{\partial \bar{z}^{\gamma}} \tag{8.24}$$

and the covariant derivative of a vector ξ^i, say, is given by

$$\left\{ \begin{array}{ll} \xi^{\alpha}{}_{;\gamma} = \dfrac{\partial \xi^{\alpha}}{\partial z^{\gamma}} + \Gamma^{\alpha}_{\beta\gamma}\xi^{\beta} , & \xi^{\bar{\alpha}}{}_{;\gamma} = \dfrac{\partial \xi^{\bar{\alpha}}}{\partial z^{\gamma}} , \\ & \\ \xi^{\alpha}{}_{;\bar{\gamma}} = \dfrac{\partial \xi^{\alpha}}{\partial \bar{z}^{\gamma}} , & \xi^{\bar{\alpha}}{}_{;\bar{\gamma}} = \dfrac{\partial \xi^{\bar{\alpha}}}{\partial \bar{z}^{\gamma}} + \Gamma^{\bar{\alpha}}_{\bar{\beta}\bar{\gamma}}\xi^{\bar{\beta}} . \end{array} \right. \tag{8.25}$$

2. CURVATURE IN KAEHLER MANIFOLD

From the definition of the curvature tensor

$$R^i{}_{jkl} = \frac{\partial \Gamma^i_{jk}}{\partial z^l} - \frac{\partial \Gamma^i_{jl}}{\partial z^k} + \Gamma^s_{jk}\Gamma^i_{sl} - \Gamma^s_{jl}\Gamma^i_{sk}$$

we obtain

$$(8.26) \qquad R^\alpha{}_{\bar\beta kl} = 0 \quad \text{and} \quad R^{\bar\alpha}{}_{\beta kl} = 0$$

and for

$$R_{ijkl} = g_{is}R^s{}_{jkl}$$

we have

$$(8.27) \qquad R_{\alpha\beta kl} = 0 \quad \text{and} \quad R_{\bar\alpha\bar\beta kl} = 0$$

Also,

$$R_{ijkl} = R_{klij}$$

implies

$$(8.28) \qquad R_{ij\gamma\delta} = 0 \quad \text{and} \quad R_{ij\bar\gamma\bar\delta} = 0$$

From (8.27) and (8.28), we can see that only the components of the form

$$R_{\alpha\bar\beta\gamma\bar\delta}\,, \quad R_{\alpha\bar\beta\bar\gamma\delta}\,, \quad R_{\bar\alpha\beta\gamma\bar\delta}\,, \quad R_{\bar\alpha\beta\bar\gamma\delta}$$

can be different from zero, and consequently that only the components of the form

$$R^\alpha{}_{\beta\gamma\bar\delta}\,, \quad R^\alpha{}_{\beta\bar\gamma\delta}\,, \quad R^{\bar\alpha}{}_{\bar\beta\gamma\bar\delta}\,, \quad R^{\bar\alpha}{}_{\bar\beta\bar\gamma\delta}$$

can be different from zero, and also we obtain

$$(8.29) \qquad R^\alpha{}_{\beta\gamma\bar\delta} = \frac{\partial \Gamma^\alpha_{\beta\gamma}}{\partial \bar z^\delta}$$

This equation shows that if the components $\Gamma^\alpha_{\beta\gamma}$ are complex analytic functions of z^α, then all components $R^i{}_{jkl}$ of the curvature tensor vanish.

From the Bianchi identity:

$$R^i{}_{jkl} + R^i{}_{klj} + R^i{}_{ljk} = 0$$

we have

$$R^\alpha{}_{\beta\gamma\bar\delta} + R^\alpha{}_{\gamma\bar\delta\beta} + R^\alpha{}_{\bar\delta\beta\gamma} = 0$$

But, the last term of the left-hand member being zero, we have

$$(8.30) \qquad \begin{cases} R^\alpha{}_{\beta\bar\delta\gamma} = R^\alpha{}_{\gamma\bar\delta\beta} \\ R^\alpha{}_{\beta\gamma\bar\delta} = R^\alpha{}_{\gamma\beta\bar\delta} \end{cases}$$

and this can also be obtained directly from (8.29).

Next, from the definition of R_{ijkl}, we have

$$R_{\alpha\bar\beta\gamma\bar\delta} = g_{\alpha\bar\epsilon} R^{\bar\epsilon}{}_{\bar\beta\gamma\bar\delta} = - g_{\alpha\bar\epsilon} \frac{\partial \Gamma^{\bar\epsilon}_{\bar\beta\bar\delta}}{\partial z^\gamma}$$

$$= - g_{\alpha\bar\epsilon} \frac{\partial}{\partial z^\gamma} \left(g^{\bar\epsilon\tau} \frac{\partial g_{\tau\bar\beta}}{\partial \bar z^\delta} \right)$$

and hence

$$(8.31) \qquad R_{\alpha\bar\beta\gamma\bar\delta} = - \frac{\partial^2 g_{\alpha\bar\beta}}{\partial z^\gamma \partial \bar z^\delta} + g^{\bar\epsilon\tau} \frac{\partial g_{\alpha\bar\epsilon}}{\partial z^\gamma} \frac{\partial g_{\bar\beta\tau}}{\partial \bar z^\delta}$$

and also

$$(8.32) \qquad R_{\alpha\bar\beta\gamma\bar\delta} = - \frac{\partial^4 \Phi}{\partial z^\alpha \partial \bar z^\beta \partial z^\gamma \partial \bar z^\delta} + g^{\bar\epsilon\tau} \frac{\partial^3 \Phi}{\partial \bar z^\epsilon \partial z^\alpha \partial z^\gamma} \frac{\partial^3 \Phi}{\partial z^\tau \partial \bar z^\beta \partial \bar z^\delta}$$

and the latter implies

$$(8.33) \qquad R_{\alpha\bar\beta\gamma\bar\delta} = R_{\gamma\bar\beta\alpha\bar\delta} = R_{\alpha\bar\delta\gamma\bar\beta} = R_{\gamma\bar\delta\alpha\bar\beta}$$

For the Ricci tensor R_{ij}, we have

(8.34) $$R_{\beta\gamma} = R^{i}{}_{\beta\gamma i} = R^{\alpha}{}_{\beta\gamma\alpha} + R^{\bar{\alpha}}{}_{\beta\gamma\bar{\alpha}} = 0$$

and consequently

(8.35) $$R_{\bar{\beta}\bar{\gamma}} = 0$$

and

$$R_{\beta\bar{\gamma}} = R^{\alpha}{}_{\beta\bar{\gamma}\alpha} = - R^{\alpha}{}_{\beta\alpha\bar{\gamma}} = - \frac{\partial\Gamma^{\alpha}_{\beta\alpha}}{\partial\bar{z}^{\gamma}}$$

But, since

$$\Gamma^{\alpha}_{\beta\alpha} = \frac{\partial \log\sqrt{g}}{\partial z^{\beta}}$$

we have

(8.36) $$R_{\beta\bar{\gamma}} = - \frac{\partial^{2}\log\sqrt{g}}{\partial z^{\beta}\partial\bar{z}^{\gamma}}$$

where

$$g = |g_{ij}| = |g_{\alpha\bar{\beta}}|^{2}$$

We now introduce a sectional curvature K determined by two linearly independent vectors u^{i} and v^{i} :

(8.37) $$K = \frac{R_{ijkl}u^{i}v^{j}u^{k}v^{l}}{(g_{jk}g_{il} - g_{jl}g_{ik})u^{i}v^{j}u^{k}v^{l}}$$

If this sectional curvature is the same for all possible 2-dimensional section, then the curvature tensor must have the form

(8.38) $$R_{ijkl} = K(g_{jk}g_{il} - g_{jl}g_{ik})$$

but in the present case this reduces to

$$R_{\alpha\bar{\beta}\gamma\bar{\delta}} = Kg_{\bar{\beta}\gamma}g_{\alpha\bar{\delta}}$$

and on substituting this into

$$R_{\alpha\bar{\beta}\gamma\bar{\delta}} = R_{\gamma\bar{\beta}\alpha\bar{\delta}}$$

we find

$$Kg_{\bar{\beta}\gamma}g_{\alpha\bar{\delta}} = Kg_{\bar{\beta}\alpha}g_{\gamma\bar{\delta}}$$

If we multiply this by $g^{\bar{\beta}\gamma}g^{\alpha\bar{\delta}}$ and contract, we obtain

$$n^2K = nK$$

and hence the following conclusion.

THEOREM 8.1. For $n > 1$, if, at every point of a Kaehler manifold, the sectional curvature is the same for all possible 2-dimensional sections, then the curvature tensor is identically zero.

Now, if the two vectors u^i and v^i satisfy the conditions:

$$v^{\alpha} = iu^{\alpha}, \qquad v^{\bar{\alpha}} = -iu^{\bar{\alpha}} \tag{8.39}$$

then the section is called a <u>holomorphic</u> section. For a holomorphic section, we have

$$R_{ijkl}u^iv^ju^kv^l = -4R_{\alpha\bar{\beta}\gamma\bar{\delta}}u^{\alpha}u^{\bar{\beta}}u^{\gamma}u^{\bar{\delta}}$$

$$(g_{jk}g_{il} - g_{jl}g_{ik})u^iv^ju^kv^l = -4g_{\alpha\bar{\beta}}g_{\gamma\bar{\delta}}u^{\alpha}u^{\bar{\beta}}u^{\gamma}u^{\bar{\delta}}$$

and consequently

$$K = \frac{R_{\alpha\bar{\beta}\gamma\bar{\delta}}u^{\alpha}u^{\bar{\beta}}u^{\gamma}u^{\bar{\delta}}}{g_{\alpha\bar{\beta}}g_{\gamma\bar{\delta}}u^{\alpha}u^{\bar{\beta}}u^{\gamma}u^{\bar{\delta}}} \tag{8.40}$$

$$= \frac{2R_{\alpha\bar{\beta}\gamma\bar{\delta}}u^{\alpha}u^{\bar{\beta}}u^{\gamma}u^{\bar{\delta}}}{(g_{\alpha\bar{\beta}}g_{\gamma\bar{\delta}} + g_{\alpha\bar{\delta}}g_{\gamma\bar{\beta}})u^{\alpha}u^{\bar{\beta}}u^{\gamma}u^{\bar{\delta}}}$$

Thus, if we assume that at all points of the manifold, the holomorphic sectional curvature are all the same, then we must have

$$[R_{\alpha\bar{\beta}\gamma\bar{\delta}} - \frac{K}{2}(g_{\alpha\bar{\beta}}g_{\gamma\bar{\delta}} + g_{\alpha\bar{\delta}}g_{\gamma\bar{\beta}})]u^{\alpha}u^{\bar{\beta}}u^{\gamma}u^{\bar{\delta}} = 0$$

for any u^α, from which

(8.41) $$R_{\alpha\bar{\beta}\gamma\bar{\delta}} = \frac{K}{2}(g_{\alpha\bar{\beta}}g_{\gamma\bar{\delta}} + g_{\alpha\bar{\delta}}g_{\gamma\bar{\beta}})$$

On the other hand, from the Bianchi identity:

$$R_{ijkl;m} + R_{ijlm;k} + R_{ijmk;l} = 0$$

we obtain

$$R_{\alpha\bar{\beta}\gamma\bar{\delta};\epsilon} + R_{\alpha\bar{\beta}\bar{\delta}\epsilon;\gamma} + R_{\alpha\bar{\beta}\epsilon\gamma;\bar{\delta}} = 0$$

or

(8.42) $$R_{\alpha\bar{\beta}\gamma\bar{\delta};\epsilon} = R_{\alpha\bar{\beta}\epsilon\bar{\delta};\gamma}$$

Substituting (8.41) into (8.42), we find

$$K_{;\epsilon}(g_{\alpha\bar{\beta}}g_{\gamma\bar{\delta}} + g_{\alpha\bar{\delta}}g_{\gamma\bar{\beta}}) = K_{;\gamma}(g_{\alpha\bar{\beta}}g_{\epsilon\bar{\delta}} + g_{\alpha\bar{\delta}}g_{\epsilon\bar{\beta}})$$

and contracting with $g^{\alpha\bar{\beta}}g^{\gamma\bar{\delta}}$, we obtain

$$n(n+1)K_{;\epsilon} = (n+1)K_{;\epsilon}$$

hence, for $n > 1$,

$$K_{;\epsilon} = 0$$

and

$$K_{;\bar{\epsilon}} = 0$$

Hence

THEOREM 8.2. If, at all points of a Kaehler manifold, the holomorphic sectional curvature K is the same, then the curvature tensor has the form (8.41) and K is an absolute constant.

We shall call such a manifold, manifold of constant holomorphic curvature.

THEOREM 8.3. In a manifold of constant holomorphic curvature k, for the general sectional curvature K, we have

$$0 < \frac{1}{4} k \leq K \leq k \qquad \text{if } k > 0 \tag{8.43}$$

and

$$k \leq K \leq \frac{1}{4} k < 0 \qquad \text{if } k < 0 \tag{8.44}$$

where the upper limit in the first case (lower limit in the second case) is attained when the section is holomorphic and the lower limit in the first case (upper limit in the second case) is attained when the inner product of two vectors defining the section is real valued. (Bochner [3]).

In fact, we have

$$\begin{aligned} R_{ijkl}u^i v^j u^k v^l &= R_{\alpha\bar{\beta}\gamma\bar{\delta}}(u^\alpha v^{\bar{\beta}} - u^{\bar{\beta}} v^\alpha)(u^\gamma v^{\bar{\delta}} - u^{\bar{\delta}} v^\gamma) \\ &= \frac{k}{2}(g_{\alpha\bar{\beta}} g_{\gamma\bar{\delta}} + g_{\alpha\bar{\delta}} g_{\gamma\bar{\beta}})(u^\alpha v^{\bar{\beta}} - u^{\bar{\beta}} v^\alpha)(u^\gamma v^{\bar{\delta}} - u^{\bar{\delta}} v^\gamma) \\ &= k[(uv)^2 + (vu)^2 - (uv)(vu) - (uu)(vv)] \end{aligned}$$

and

$$\begin{aligned} (g_{jk}g_{il} - g_{jl}g_{ik})u^i v^j u^k v^l &= [g_{\alpha\bar{\beta}}u^\alpha v^{\bar{\beta}} + g_{\alpha\bar{\beta}}v^\alpha u^{\bar{\beta}}]^2 - 4[g_{\alpha\bar{\beta}}u^\alpha u^{\bar{\beta}}][g_{\gamma\bar{\delta}}v^\gamma v^{\bar{\delta}}] \\ &= (uv)^2 + (vu)^2 + 2(uv)(vu) - 4(uu)(vv) \end{aligned}$$

where we have put

$$(uv) = g_{\alpha\bar{\beta}}u^\alpha v^{\bar{\beta}}$$

and thus

$$K = k\,\frac{(uv)^2 + (vu)^2 - (uv)(vu) - (uu)(vv)}{(uv)^2 + (vu)^2 + 2(uv)(vu) - 4(uu)(vv)}$$

Now, if we put

$$\frac{(uv)}{\sqrt{(uu)(vv)}} = re^{+i\theta} \qquad (0 \leq r;\ r,\ \theta;\ \text{real})$$

then

$$\frac{(vu)}{\sqrt{(uu)(vv)}} = re^{-i\theta}$$

and

$$0 \leq r \leq 1$$

and

$$K = k\,\frac{1 + r^2 - 2r^2\cos 2\theta}{4 - 2r^2 - 2r^2\cos 2\theta}$$

$$= k\left[1 - \frac{3}{4}\,\frac{1 - r^2}{1 - r^2\cos^2\theta}\right]$$

Since we have

$$0 \leq \frac{1 - r^2}{1 - r^2\cos^2\theta} \leq 1$$

we can conclude that, for $k > 0$ we have (8.43) and similarly for (8.44).

The Ricci curvature is given by

$$\frac{R_{ij}u^i u^j}{g_{ij}u^i u^j} = \frac{R_{\alpha\bar{\beta}}u^\alpha u^{\bar{\beta}}}{g_{\alpha\bar{\beta}}u^\alpha u^{\bar{\beta}}}$$

and for the manifold of constant holomorphic curvature, we have, from (8.41),

$$R_{\alpha\bar{\beta}} = R_{\bar{\beta}\alpha} = \frac{n+1}{2}Kg_{\alpha\bar{\beta}} \tag{8.45}$$

and consequently

$$\frac{R_{\alpha\bar{\beta}}u^\alpha u^{\bar{\beta}}}{g_{\alpha\bar{\beta}}u^\alpha u^{\bar{\beta}}} = \frac{n+1}{2}K$$

is constant.

In general, if the Ricci tensor $R_{\alpha\bar{\beta}}$ satisfies

$$R_{\alpha\bar{\beta}} = \lambda g_{\alpha\bar{\beta}} \tag{8.46}$$

then

$$\lambda = \frac{1}{n} g^{\alpha\bar{\beta}} R_{\alpha\bar{\beta}} = \frac{1}{2n} R$$

and from the identity:

$$R^{i}{}_{j;i} = \frac{1}{2} R_{;j}$$

we get

$$R^{\alpha}{}_{\beta;\alpha} = \frac{1}{2} R_{;\beta}$$

Substituting (8.46) into this, we get

$$\frac{1}{2n} R_{;\beta} = \frac{1}{2} R_{;\beta}$$

from which $R_{;\beta} = 0$; and similarly $R_{;\bar{\beta}} = 0$.

THEOREM 8.4. In an Einstein manifold:

$$R_{\alpha\bar{\beta}} = \lambda g_{\alpha\bar{\beta}}$$

λ is an absolute constant.

3. COVARIANT AND CONTRAVARIANT ANALYTIC VECTOR FIELDS

If the components ξ_{α} of a covariant vector field are complex analytic functions of the coordinates (z^{α}) , then we have

$$\xi_{\alpha;\bar{\beta}} = 0 \tag{8.47}$$

therefore from the Ricci identity:

$$\xi_{\beta;\gamma;\bar{\delta}} - \xi_{\beta;\bar{\delta};\gamma} = - \xi_{\alpha} R^{\alpha}{}_{\beta\gamma\bar{\delta}}$$

we obtain

$$\xi_{\beta;\gamma;\bar{\delta}} - \xi_\alpha R^\alpha{}_{\gamma\bar{\delta}\beta} = 0$$

and thus also

$$g^{\gamma\bar{\delta}}\xi_{\beta;\gamma;\bar{\delta}} - \xi_\alpha R^\alpha{}_\beta = 0 \tag{8.48}$$

Next, if $\phi(z, \bar{z})$ is real valued, then

$$\Delta\phi = g^{ij}\phi_{;i;j} = g^{\alpha\bar{\beta}}\phi_{;\alpha;\bar{\beta}} + g^{\bar{\alpha}\beta}\phi_{;\bar{\alpha};\beta}$$

or

$$\Delta\phi = 2g^{\alpha\bar{\beta}} \frac{\partial^2\phi}{\partial z^\alpha \partial \bar{z}^\beta} \tag{8.49}$$

Now, if we put

$$z^\alpha = x^\alpha + ix^{\bar{\alpha}} \qquad \bar{z}^\alpha = x^\alpha - ix^{\bar{\alpha}}$$

then $\Delta\phi$ will take the form of Laplacean in real variables $(x^\alpha, x^{\bar{\alpha}})$, and consequently, Theorem 2.3 of Hopf-Bochner applies again.

Now, if the components ξ_α of a self-adjoint vector field are complex analytic, then for

$$\phi = 2g^{\alpha\bar{\beta}}\xi_\alpha\xi_{\bar{\beta}}$$

we have

$$\Delta\phi = 4g^{\gamma\bar{\delta}}(g^{\alpha\bar{\beta}}\xi_\alpha\xi_{\bar{\beta}})_{;\gamma;\bar{\delta}}$$

$$= 4[g^{\alpha\bar{\beta}}g^{\gamma\bar{\delta}}\xi_{\alpha;\gamma}\xi_{\bar{\beta};\bar{\delta}} + g^{\alpha\bar{\beta}}g^{\gamma\bar{\delta}}\xi_{\alpha;\gamma;\bar{\delta}}\xi_{\bar{\beta}}]$$

and (8.48) implies

$$\Delta\phi = 4[g^{\alpha\bar{\beta}}g^{\gamma\bar{\delta}}\xi_{\alpha;\gamma}\xi_{\bar{\beta};\bar{\delta}} + R_{\alpha\bar{\beta}}\xi^\alpha\xi^{\bar{\beta}}] \tag{8.50}$$

Thus we have

THEOREM 8.5. In a compact Kaehler manifold with $R_{\alpha\bar{\beta}}\xi^\alpha\xi^{\bar{\beta}} \geq 0$, a self-adjoint covariant vector field whose components are analytic functions of coordinates must have vanishing covariant derivative, and if $R_{\alpha\bar{\beta}}\xi^{\alpha}\xi^{\bar{\beta}}$ is positive definite, no such covariant vector field exists other than zero. (Bochner [2]).

For a contravariant ξ^{α} with complex analytic components, we have

$$\xi^{\alpha}{}_{;\bar{\gamma}} = 0 \tag{8.51}$$

and the Ricci identity:

$$\xi^{\alpha}{}_{;\gamma;\bar{\delta}} - \xi^{\alpha}{}_{;\bar{\delta};\gamma} = \xi^{\beta}R^{\alpha}{}_{\beta\gamma\bar{\delta}}$$

implies

$$\xi^{\alpha}{}_{;\gamma;\bar{\delta}} + R^{\alpha}{}_{\gamma\bar{\delta}\beta}\xi^{\beta} = 0$$

and hence

$$g^{\gamma\bar{\delta}}\xi^{\alpha}{}_{;\gamma;\bar{\delta}} + R^{\alpha}{}_{\beta}\xi^{\beta} = 0 \tag{8.52}$$

We now have

$$\begin{aligned}\Delta\phi &= 4g^{\gamma\bar{\delta}}(g_{\alpha\bar{\beta}}\xi^{\alpha}\xi^{\bar{\beta}})_{;\gamma;\bar{\delta}} \\ &= 4[g_{\alpha\bar{\beta}}g^{\gamma\bar{\delta}}\xi^{\alpha}{}_{;\gamma}\xi^{\bar{\beta}}{}_{;\bar{\delta}} + g_{\alpha\bar{\beta}}(g^{\gamma\bar{\delta}}\xi^{\alpha}{}_{;\gamma;\bar{\delta}})\xi^{\bar{\beta}}]\end{aligned}$$

and on substituting from (8.52), we find

$$\Delta\phi = 4[g_{\alpha\bar{\beta}}g^{\gamma\bar{\delta}}\xi^{\alpha}{}_{;\gamma}\xi^{\bar{\beta}}{}_{;\bar{\delta}} - R_{\alpha\bar{\beta}}\xi^{\alpha}\xi^{\bar{\beta}}] \tag{8.53}$$

and hence the following conclusion:

THEOREM 8.6. In a compact Kaehler manifold with $R_{\alpha\bar{\beta}}\xi^{\alpha}\xi^{\bar{\beta}} \leq 0$, a self-adjoint contravariant vector field whose components are analytic functions of coordinates must have vanishing covariant derivative, and if $R_{\alpha\bar{\beta}}\xi^{\alpha}\xi^{\bar{\beta}}$ is negative definite, no such contravariant vector field exists other than zero. (Bochner [2]).

4. COMPLEX ANALYTIC MANIFOLDS ADMITTING A TRANSITIVE COMMUTATIVE GROUP OF TRANSFORMATIONS

We consider a complex analytic manifold of real dimension $2n$ which admits a transitive commutative group of transformations whose infinitesimal operators are

$$X_p f = \eta_p^\alpha \frac{\partial f}{\partial z^\alpha} \tag{8.54}$$

where η_p^α $(p, q, \ldots = 1, 2, \ldots, r)$ are r holomorphic contravariant vector fields. By the transivity of the group, we have $r \geq n$ and the rank of the matrix (η_p^α) is n. By the commutativity of the group, we have

$$\phi_{pq}{}^\alpha \equiv \eta_p^\beta \frac{\partial \eta_q^\alpha}{\partial z^\beta} - \eta_q^\beta \frac{\partial \eta_p^\alpha}{\partial z^\beta} = 0 \tag{8.55}$$

Now, if we introduce the Hermitian tensor field

$$g^{\alpha\bar{\beta}} = \sum_{p=1}^{r} \eta_p^\alpha \overline{\eta_p^\beta} \tag{8.56}$$

then, due to the transitivity postulated, it is strictly positive definite and thus has the inverse $g_{\alpha\bar{\beta}}$.

On the other hand, multiplying (8.55) by $\overline{\eta_p^\rho}\,\overline{\eta_q^\sigma}$ and summing up with respect to p and q, we get

$$g^{\beta\bar{\rho}} \frac{\partial g^{\alpha\bar{\sigma}}}{\partial z^\beta} - g^{\beta\bar{\sigma}} \frac{\partial g^{\alpha\bar{\rho}}}{\partial z^\beta} = 0$$

Multiplying this by $g_{\alpha\bar{\gamma}}$ and summing up with respect to α, we find

$$g^{\alpha\bar{\sigma}} g^{\beta\bar{\rho}} \left(\frac{\partial g_{\alpha\bar{\gamma}}}{\partial z^\beta} - \frac{\partial g_{\beta\bar{\gamma}}}{\partial z^\alpha} \right) = 0$$

from which

$$\frac{\partial g_{\alpha\bar{\gamma}}}{\partial z^\beta} = \frac{\partial g_{\beta\bar{\gamma}}}{\partial z^\alpha}$$

thus, $g_{\alpha\bar{\beta}}$ satisfies the Kaehler's condition.

Hence, the $\Gamma^{\alpha}_{\beta\gamma}$ having the form

$$\Gamma^{\alpha}_{\beta\gamma} = g^{\alpha\bar{\epsilon}} \frac{\partial g_{\bar{\epsilon}\gamma}}{\partial z^{\beta}} = - \frac{\partial g^{\alpha\bar{\epsilon}}}{\partial z^{\beta}} g_{\bar{\epsilon}\gamma}$$

we have

$$\eta^{\alpha}_{p;\gamma} = \frac{\partial \eta^{\alpha}_{p}}{\partial z^{\gamma}} + \eta^{\beta}_{p}\Gamma^{\alpha}_{\beta\gamma} = \frac{\partial \eta^{\alpha}_{p}}{\partial z^{\gamma}} - \eta^{\beta}_{p} \frac{\partial g^{\alpha\bar{\epsilon}}}{\partial z^{\beta}} g_{\bar{\epsilon}\gamma}$$

But, on the other hand, multiplying (8.55) by $\overline{\eta^{\epsilon}_{q}}$ and summing up with respect to q , we find

$$\eta^{\beta}_{p} \frac{\partial g^{\alpha\bar{\epsilon}}}{\partial z^{\beta}} - g^{\beta\bar{\epsilon}} \frac{\partial \eta^{\alpha}_{p}}{\partial z^{\beta}} = 0$$

or

$$\eta^{\beta}_{p} \frac{\partial g^{\alpha\bar{\epsilon}}}{\partial z^{\beta}} g_{\bar{\epsilon}\gamma} - \frac{\partial \eta^{\alpha}_{p}}{\partial z^{\gamma}} = 0$$

which shows that

$$\eta^{\alpha}_{p;\gamma} = 0$$

Thus the vector fields η^{α}_{p} are parallel in the constructed metric.

Now, this implies that in the Ricci identity:

$$\eta^{\alpha}_{p;\gamma;\bar{\delta}} - \eta^{\alpha}_{p;\bar{\delta};\gamma} = \eta^{\beta}_{p}R^{\alpha}{}_{\beta\gamma\bar{\delta}}$$

the left side vanishes identically, and hence also the right side. But η^{α}_{p} has everywhere maximal rank, so that

$$R^{\alpha}{}_{\beta\gamma\bar{\delta}} = 0$$

and thus our manifold is a flat Kaehler manifold. Therefore, in the neighborhood of every point we can allowably normalize the metric tensor to

$$g_{\alpha\bar{\beta}} = \delta_{\alpha\bar{\beta}}$$

This normalization shows that the covariant vector fields

$$\eta_\alpha^p = g_{\alpha\bar\beta}\,\overline{\eta_p^\beta}$$

are likewise holomorphic and parallel, so that in particular we have

$$\frac{\partial\eta_\alpha^p}{\partial z^\beta} - \frac{\partial\eta_\beta^p}{\partial z^\alpha} = 0$$

Therefore the r Abelian integrals

$$w^p(z) = \int_{z_0}^{z} \eta_\alpha^p dz^\alpha$$

may be introduced, and if, for instance, the first n among them are linearly independent at the point z_0, then they will be so everywhere and they map the manifold holomorphically and locally one-to-one into the Euclidean manifold.

THEOREM 8.7. Let V_{2n}, $n \geq 1$, be a complex analytic manifold of real dimension $2n$. If, for $r \geq n$, there are on it r holomorphic contravariant vector fields η_p^α such that its rank has everywhere its maximal value n and that the bracket expressions (8.55) all vanish identically, then there are on the manifold n simple Abelian integrals of the first kind by which it is mapped holomorphically and locally one-to-one into the Euclidean manifold. In particular, if V_{2n} is compact, it is a complex multi-torus. (Bochner [9]).

5. SELF-ADJOINT VECTOR SATISFYING $\xi_{\bar\alpha} = \partial f/\partial\bar z^\alpha$ AND $\Delta f = 0$

We now consider a (self-adjoint) vector field which in the neighborhood of every point can be represented in the form

$$(8.57) \qquad \xi_\alpha = \frac{\partial f}{\partial z^\alpha} \qquad\qquad \xi_{\bar\alpha} = \overline{\xi_\alpha} = \frac{\partial\bar f}{\partial\bar z^\alpha}$$

It is not a local gradient field in the proper sense unless

$f = \bar{f}$, that is, if f is real valued; and in our application it will definitely not be so.

We introduce the associate vector

$$\eta_\alpha = \frac{\partial \bar{f}}{\partial z^\alpha} \qquad \eta_{\bar{\alpha}} = \frac{\partial f}{\partial \bar{z}^\alpha} \tag{8.58}$$

and these two vectors have the following properties:

$$\xi_{\alpha;\beta} = \xi_{\beta;\alpha}, \qquad \xi_{\bar{\alpha};\bar{\beta}} = \xi_{\bar{\beta};\bar{\alpha}},$$

$$\eta_{\alpha;\beta} = \eta_{\beta;\alpha}, \qquad \eta_{\bar{\alpha};\bar{\beta}} = \eta_{\bar{\beta};\bar{\alpha}},$$

$$\xi_{\alpha;\bar{\beta}} = \eta_{\bar{\beta};\alpha}, \qquad \xi_{\bar{\alpha};\beta} = \eta_{\beta;\bar{\alpha}}.$$

Now, for $\phi = 2g^{\alpha\bar{\beta}}\xi_\alpha\xi_{\bar{\beta}}$, we obtain

$$\frac{1}{4}\Delta\phi = A + B + C$$

where

$$A = g^{\alpha\bar{\beta}}g^{\rho\bar{\sigma}}(\xi_{\alpha;\rho}\xi_{\bar{\beta};\bar{\sigma}} + \xi_{\alpha;\bar{\sigma}}\xi_{\bar{\beta};\rho})$$

$$B = g^{\alpha\bar{\beta}}g^{\rho\bar{\sigma}}\xi_{\alpha;\rho;\bar{\sigma}}\xi_{\bar{\beta}}$$

$$C = g^{\alpha\bar{\beta}}g^{\rho\bar{\sigma}}\xi_\alpha\xi_{\bar{\beta};\rho;\bar{\sigma}}$$

If we substitute

$$\xi_{\alpha;\rho;\bar{\sigma}} = \xi_{\rho;\alpha;\bar{\sigma}} = (\xi_{\rho;\alpha;\bar{\sigma}} - \xi_{\rho;\bar{\sigma};\alpha}) + \xi_{\rho;\bar{\sigma};\alpha}$$

$$= -\xi_\lambda R^\lambda{}_{\rho\alpha\bar{\sigma}} + \xi_{\rho;\bar{\sigma};\alpha}$$

we obtain

$$B = R_{\alpha\bar{\beta}}\xi^\alpha\xi^{\bar{\beta}} + g^{\alpha\bar{\beta}}(g^{\rho\bar{\sigma}}\xi_{\rho;\bar{\sigma}})_{;\alpha}\xi_{\bar{\beta}}$$

and if we put

$$\xi_{\bar\beta;\rho;\bar\sigma} = \eta_{\rho;\bar\beta;\bar\sigma} = \eta_{\rho;\bar\sigma;\bar\beta} + (\eta_{\rho;\bar\beta;\bar\sigma} - \eta_{\rho;\bar\sigma;\bar\beta})$$

$$= \eta_{\rho;\bar\sigma;\bar\beta} - \eta_\lambda R^\lambda{}_{\rho\bar\beta\bar\sigma}$$

$$= \eta_{\rho;\bar\sigma;\bar\beta}$$

we obtain

$$C = g^{\alpha\bar\beta}(g^{\rho\bar\sigma}\eta_{\rho;\bar\sigma})_{;\bar\beta}\xi_\alpha$$

Finally we introduce the assumption

(8.59) $$g^{\rho\bar\sigma}\xi_{\rho;\bar\sigma} = 0$$

that is

(8.60) $$\Delta f = 0$$

This will also imply

$$g^{\rho\bar\sigma}\eta_{\rho;\bar\sigma} = 0$$

and if we interchange the variables (z^α) and $(\bar z^\alpha)$, we obtain the following theorem:

THEOREM 8.8. If on a compact manifold with positive Ricci curvature a (self-adjoint) vector field ξ_i has the property that in the neighborhood of every point the components $\xi_{\bar\alpha}$ can be expressed in the form

(8.61) $$\xi_{\bar\alpha} = \frac{\partial f}{\partial \bar z^\alpha} \quad \text{with} \quad \Delta f = 0$$

then $\xi_{\bar\alpha} = 0$, that is, f is complex analytic.

If the Ricci curvature is only non negative, then $\xi_{\bar\alpha;i} = 0$, that is, the derivatives $\partial f/\partial \bar z^\alpha$ are not necessarily zero but have covariant derivative zero. (Bochner [2]).

6. ANALYTIC TENSORS

If the components

$$\xi^{\alpha_1\alpha_2\cdots\alpha_p}{}_{\beta_1\beta_2\cdots\beta_q}$$

of a self-adjoint tensor of mixed type are complex analytic functions of the coordinates (z^α), then we again have

$$\xi^{\alpha_1\alpha_2\cdots\alpha_p}{}_{\beta_1\beta_2\cdots\beta_q;\bar{\gamma}} = 0 \tag{8.62}$$

and from the Ricci identity:

$$\xi^{\alpha_1\alpha_2\cdots\alpha_p}{}_{\beta_1\beta_2\cdots\beta_q;\gamma;\bar{\delta}} - \xi^{\alpha_1\alpha_2\cdots\alpha_p}{}_{\beta_1\beta_2\cdots\beta_q;\bar{\delta};\gamma}$$

$$= \xi^{\lambda\alpha_2\cdots\alpha_p}{}_{\beta_1\beta_2\cdots\beta_q}R^{\alpha_1}{}_{\lambda\gamma\bar{\delta}} + \cdots + \xi^{\alpha_1\alpha_2\cdots\alpha_{p-1}\lambda}{}_{\beta_1\beta_2\cdots\beta_q}R^{\alpha_p}{}_{\lambda\gamma\bar{\delta}}$$

$$- \xi^{\alpha_1\alpha_2\cdots\alpha_p}{}_{\mu\beta_2\cdots\beta_q}R^{\mu}{}_{\beta_1\gamma\bar{\delta}} - \cdots - \xi^{\alpha_1\alpha_2\cdots\alpha_p}{}_{\beta_1\beta_2\cdots\beta_{q-1}\mu}R^{\mu}{}_{\beta_q\gamma\bar{\delta}}$$

by using (8.62) and contracting with $g^{\gamma\bar{\delta}}$, we obtain

$$g^{\gamma\bar{\delta}}\xi^{\alpha_1\alpha_2\cdots\alpha_p}{}_{\beta_1\beta_2\cdots\beta_q;\gamma;\bar{\delta}} = -\xi^{\lambda\alpha_2\cdots\alpha_p}{}_{\beta_1\beta_2\cdots\beta_q}R^{\alpha_1}{}_{\lambda} \tag{8.63}$$

$$- \cdots - \xi^{\alpha_1\alpha_2\cdots\alpha_{p-1}\lambda}{}_{\beta_1\beta_2\cdots\beta_q}R^{\alpha_p}{}_{\lambda} + \xi^{\alpha_1\alpha_2\cdots\alpha_p}{}_{\mu\beta_2\cdots\beta_q}R^{\mu}{}_{\beta_1}$$

$$+ \cdots + \xi^{\alpha_1\alpha_2\cdots\alpha_p}{}_{\beta_1\beta_2\cdots\beta_{q-1}\mu}R^{\mu}{}_{\beta_q}$$

Now, if we put

$$\Phi = g_{\alpha_1\bar{\gamma}_1}\cdots g_{\alpha_p\bar{\gamma}_p}g^{\beta_1\bar{\delta}_1}\cdots g^{\beta_q\bar{\delta}_q}\xi^{\alpha_1\cdots\alpha_p}{}_{\beta_1\cdots\beta_q}\xi^{\bar{\gamma}_1\cdots\bar{\gamma}_p}{}_{\bar{\delta}_1\cdots\bar{\delta}_q}$$

then we have

$$\Delta\phi = 2[g_{\alpha_1\bar{\gamma}_1} \cdots g_{\alpha_p\bar{\gamma}_p} g^{\beta_1\bar{\delta}_1} \cdots g^{\beta_q\bar{\delta}_q} g^{\sigma\bar{\tau}} \xi^{\alpha_1\cdots\alpha_p}{}_{\beta_1\cdots\beta_q;\sigma} \xi^{\bar{\gamma}_1\cdots\bar{\gamma}_p}{}_{\bar{\delta}_1\cdots\bar{\delta}_q;\bar{\tau}}$$

$$+ g_{\alpha_1\bar{\gamma}_1} \cdots g_{\alpha_p\bar{\gamma}_p} g^{\beta_1\bar{\delta}_1} \cdots g^{\beta_q\bar{\delta}_q} g^{\sigma\bar{\tau}} \xi^{\alpha_1\cdots\alpha_p}{}_{\beta_1\cdots\beta_q;\sigma;\bar{\tau}} \xi^{\bar{\gamma}_1\cdots\bar{\gamma}_p}{}_{\bar{\delta}_1\cdots\bar{\delta}_q}]$$

and substituting from (8.63), we have

$$\Delta\phi = 2[g_{\alpha_1\bar{\gamma}_1} \cdots g_{\alpha_p\bar{\gamma}_p} g^{\beta_1\bar{\delta}_1} \cdots g^{\beta_q\bar{\delta}_q} g^{\sigma\bar{\tau}} \xi^{\alpha_1\cdots\alpha_p}{}_{\beta_1\cdots\beta_q;\sigma} \xi^{\bar{\gamma}_1\cdots\bar{\gamma}_p}{}_{\bar{\delta}_1\cdots\bar{\delta}_q;\bar{\tau}}$$

$$+ G(\xi)]$$

where

$$(8.64) \quad G(\xi) = - \xi^{\lambda\alpha_2\cdots\alpha_p}{}_{\beta_1\cdots\beta_q} R^{\alpha_1}{}_{\lambda} \xi_{\alpha_1\cdots\alpha_p}{}^{\beta_1\cdots\beta_q}$$

$$- \cdots - \xi^{\alpha_1\cdots\alpha_{p-1}\lambda}{}_{\beta_1\cdots\beta_q} R^{\alpha_p}{}_{\lambda} \xi_{\alpha_1\cdots\alpha_p}{}^{\beta_1\cdots\beta_q}$$

$$+ \xi^{\alpha_1\alpha_2\cdots\alpha_p}{}_{\mu\beta_2\cdots\beta_q} R^{\mu}{}_{\beta_1} \xi_{\alpha_1\cdots\alpha_p}{}^{\beta_1\cdots\beta_q}$$

$$+ \cdots + \xi^{\alpha_1\cdots\alpha_p}{}_{\beta_1\cdots\beta_{q-1}\mu} R^{\mu}{}_{\beta_q} \xi_{\alpha_1\cdots\alpha_p}{}^{\beta_1\cdots\beta_q}$$

Hence the following conclusion.

THEOREM 8.9. In a compact Kaehler manifold, if the complex analytic components

$$\xi^{\alpha_1\alpha_2\cdots\alpha_p}{}_{\beta_1\beta_2\cdots\beta_q}$$

of a self-adjoint tensor of mixed type satisfy the inequality:

$$G(\xi) \geq 0$$

then we must have $G(\xi) = 0$ and

$$\xi^{\alpha_1 \ldots \alpha_p}{}_{\beta_1 \ldots \beta_q ; \tau} = 0$$

Also this assertion applies not only to tensor fields satisfying

$$\xi^{\alpha_1 \ldots \alpha_p}{}_{\beta_1 \ldots \beta_q ; \bar{\gamma}} = 0$$

but also to those satisfying

$$g^{\bar{\gamma}\delta} \xi^{\alpha_1 \ldots \alpha_p}{}_{\beta_1 \ldots \beta_q ; \bar{\gamma} ; \delta} = 0$$

(Bochner [11]).

Now, if, at every point of the manifold, we denote by M and m the algebraically largest and smallest eigenvalues of the matrix $R_{\alpha\bar{\beta}}$ respectively, then we have

$$G(\xi) \geq (qm - pM) \xi^{\alpha_1 \ldots \alpha_p}{}_{\beta_1 \ldots \beta_q} \xi_{\alpha_1 \ldots \alpha_p}{}^{\beta_1 \ldots \beta_q}$$

and we obtain the following conclusion:

THEOREM 8.10. If M and m have the meaning just stated, and if

$$qm - pM \geq 0$$

then every complex analytic tensor field of mixed type

$$\xi^{\alpha_1 \ldots \alpha_p}{}_{\beta_1 \ldots \beta_q}$$

must satisfy

$$\xi^{\alpha_1 \ldots \alpha_p}{}_{\beta_1 \ldots \beta_q ; \gamma} = 0$$

If

$$qm - pM \geq 0 \text{ everywhere and } qm - pM > 0 \text{ somewhere,}$$

then there exists no complex analytic tensor field of mixed type

$$\xi^{\alpha_1\alpha_2\cdots\alpha_p}{}_{\beta_1\beta_2\cdots\beta_q}$$

other than zero. (Bochner [11]).

As a corollary to this Theorem, we can state

THEOREM 8.11. If a compact Kaehler manifold is an Einstein manifold

$$R_{\alpha\bar{\beta}} = \lambda g_{\alpha\bar{\beta}}$$

for $\lambda > 0$, there exists no analytic tensor field of the type

$$\xi^{\alpha_1\cdots\alpha_p}{}_{\beta_1\cdots\beta_q} \quad (q > p)$$

and for $\lambda < 0$, none of the type

$$\xi^{\alpha_1\cdots\alpha_p}{}_{\beta_1\cdots\beta_q} \quad (q < p)$$

and in both cases analytic tensor field of type

$$\xi^{\alpha_1\cdots\alpha_p}{}_{\beta_1\cdots\beta_q}$$

must have vanishing covariant derivative.

Also, for $\lambda = 0$, every analytic tensor must have vanishing covariant derivative. (Bochner [11]).

7. HARMONIC VECTOR FIELDS

In a Kaehler manifold, if a vector field ξ_i (not necessarily self-adjoint) is harmonic, that is to say, if ξ_i satisfies

(8.65) $$\xi_{i;j} = \xi_{j;i} \qquad g^{ij}\xi_{i;j} = 0$$

then, by exactly the same formal calculation as in real case, we can see that ξ_i satisfies

(8.66) $$g^{jk}\xi_{i;j;k} - \xi_a R^a{}_i = 0$$

Conversely, in a compact Kaehler manifold, if a vector field (not necessarily self-adjoint) satisfies (8.66) then it is again harmonic. In fact if we make the formal transformation of coordinates

$$z^\alpha = x^\alpha + ix^{\bar{\alpha}} \qquad \bar{z}^\alpha = x^\alpha - ix^{\bar{\alpha}}$$

or

$$x^\alpha = \frac{1}{2}(z^\alpha + z^{\bar{\alpha}}) \qquad x^{\bar{\alpha}} = \frac{1}{2i}(z^\alpha - z^{\bar{\alpha}})$$

then equation (8.66) becomes

(8.67) $$g'^{jk}\xi'_{i;j;k} - \xi'_a R'^a{}_i = 0$$

and herein g'_{ij}, g'^{jk}, $\{{i \atop jk}\}'$ and $R'^a{}_i$ are all real valued. Thus, equation (8.67) implies that both the real part and the imaginary part of ξ'_i satisfy equation (8.67) and consequently, they are both harmonic vectors. Hence,

$$\xi'_{i;j} = \xi'_{j;i} \qquad g'^{ij}\xi'_{i;j} = 0$$

and also (8.65) in the original coordinate system as claimed.

THEOREM 8.12. In a compact Kaehler manifold, if a (self-adjoint or not) vector field $\xi_i = (\xi_\alpha, \xi_{\bar{\alpha}})$ is harmonic, then the vector fields

(8.68) $$\eta_i = (\xi_\alpha, 0) \qquad \zeta_i = (0, \xi_{\bar{\alpha}})$$

and the adjoint vector field

(8.69) $$(\overline{\xi_{\bar{\alpha}}}, \overline{\xi_\alpha})$$

are all three harmonic.

PROOF. Relation (8.66) splits into

$$g^{jk}\xi_{\alpha;j;k} - \xi_\beta R^\beta{}_\alpha = 0$$

$$g^{jk}\xi_{\bar\alpha;j;k} - \xi_{\bar\beta} R^{\bar\beta}{}_{\bar\alpha} = 0$$

which proves our assertion for (8.68). Also, since g_{ij}, g^{jk}, $\{{i \atop jk}\}$, and $R^a{}_i$ are self-adjoint, we get

$$g^{jk}\overline{\xi}_{\alpha;j;k} - \overline{\xi}_\beta R^{\bar\beta}{}_{\bar\alpha} = 0$$

and this proves the assertion for (8.69).

THEOREM 8.13. In a compact Kaehler manifold, a vector $\eta_i = (\xi_\alpha, 0)$ is harmonic, if and only if, all ξ_α are analytic in (z^α), and $\zeta_i = (0, \xi_{\bar\alpha})$ is harmonic, if and only if $\xi_{\bar\alpha}$ are analytic in $(\bar z^\alpha)$.

Therefore $(\xi_\alpha, \xi_{\bar\alpha})$ is harmonic, if and only if all ξ_α are analytic in (z^α) and all $\xi_{\bar\alpha}$ are analytic in $(\bar z^\alpha)$.

In fact, if η_i is harmonic, then

$$\eta_{i;j} = \eta_{j;i}$$

for $i = \alpha$, $j = \bar\beta$ implies

(8.71) $$\xi_{\alpha;\bar\beta} = 0$$

and thus ξ_α are analytic in (z^α).

Conversely, if ξ_α are analytic in (z^α), then (8.71) holds, then this and the Ricci identity:

$$\xi_{\beta;\gamma;\bar\delta} - \xi_{\beta;\bar\delta;\gamma} = -\xi_\alpha R^\alpha{}_{\beta\gamma\bar\delta}$$

imply

$$g^{\gamma\bar\delta}\xi_{\beta;\gamma;\bar\delta} - \xi_\alpha R^\alpha{}_\beta = 0$$

or

$$g^{jk}\xi_{\beta;j;k} - \xi_\alpha R^\alpha{}_\beta = 0$$

and this gives

$$g^{jk}\eta_{i;j;k} - \eta_a R^a{}_i = 0$$

so that η_i is harmonic. A similar argument applies to ζ_i, and the second half of the theorem follows now from the first half in combination with Theorem 8.12.

8. HARMONIC TENSOR FIELDS

We first of all note that on a compact Kaehler manifold, as on a real manifold, an anti-symmetric tensor field $\xi_{i_1 i_2 \ldots i_p}$ is harmonic if and only if

$$(8.72)\quad g^{jk}\xi_{i_1 i_2 \ldots i_p;j;k} - \sum_{s=1}^{p} \xi_{i_1 \ldots i_{s-1} a i_{s+1} \ldots i_p} R^a{}_{i_s}$$
$$- \sum_{s<t}^{1\ldots p} \xi_{i_1 \ldots i_{s-1} a i_{s+1} \ldots i_{t-1} b i_{t+1} \ldots i_p} R^{ab}{}_{i_s i_t} = 0$$

Now, if an anti-symmetric tensor field (not necessarily self-adjoint) satisfies (8.72), then so does every pure tensor of type h

$$(8.73)\quad p!\, \delta^{\alpha_1}_{[i_1} \ldots \delta^{\alpha_{p-h}}_{i_{p-h}} \delta^{\bar\beta_{p-h+1}}_{i_{p-h+1}} \ldots \delta^{\bar\beta_p}_{i_p]} \xi_{\alpha_1\alpha_2\ldots\alpha_{p-h}\bar\beta_{p-h+1}\ldots\bar\beta_p}$$

derived from $\xi_{i_1 i_2 \ldots i_p}$ and hence the following conclusion.

THEOREM 8.14. In a compact Kaehler manifold, if an anti-symmetric tensor

$$\xi_{i_1 i_2 \ldots i_p}$$

is harmonic, then every pure tensor (8.73) of type h (h = 0, 1, 2, ..., p) derived from

$$\xi_{i_1 i_2 \ldots i_p}$$

is also harmonic. Furthermore the adjoint tensor

$$C(\xi_{i_1 i_2 \dots i_p})$$

is also harmonic. (Hodge [1], Eckmann and Guggenheimer [1], [2], [3], [4]).

Now, if an anti-symmetric tensor field

$$\xi_{i_1 i_2 \dots i_p}$$

is harmonic, then by Theorem 8.14 the tensor field

$$\eta_{i_1 i_2 \dots i_p} = \delta_{i_1}^{\alpha_1} \delta_{i_2}^{\alpha_2} \dots \delta_{i_p}^{\alpha_p} \xi_{\alpha_1 \alpha_2 \dots \alpha_p}$$

is harmonic and we have

$$\eta_{[i_1 i_2 \dots i_p ; j]} = 0$$

from which, putting $i_1 = \alpha_1$, $i_2 = \alpha_2$, ..., $i_p = \alpha_p$ and $j = \bar{\beta}$, we get

$$\xi_{\alpha_1 \alpha_2 \dots \alpha_p ; \bar{\beta}} = 0$$

and consequently, we can conclude that

$$\xi_{\alpha_1 \alpha_2 \dots \alpha_p}$$

are complex analytic functions of coordinates (z^α).

Similarly we can prove

$$\xi_{\bar{\alpha}_1 \bar{\alpha}_2 \dots \bar{\alpha}_p ; \beta} = 0$$

and hence we have

THEOREM 8.15. In a compact Kaehler manifold, the components

$$\xi_{\alpha_1 \alpha_2 \dots \alpha_p}$$

of a harmonic tensor

$$\xi_{i_1 i_2 \ldots i_p}$$

are complex analytic functions of coordinates and the components

$$\xi_{\bar{\alpha}_1 \bar{\alpha}_2 \ldots \bar{\alpha}_p}$$

of the tensor are analytic functions of conjugate coordinates.

Conversely, if the components

$$\xi_{\alpha_1 \alpha_2 \ldots \alpha_p}$$

of an anti-symmetric tensor field

$$\xi_{i_1 i_2 \ldots i_p}$$

(not necessarily self-adjoint) are complex analytic functions of coordinates (z^α), then we have

$$\xi_{\alpha_1 \alpha_2 \ldots \alpha_p ; \bar{\beta}} = 0$$

and from the Ricci identity:

$$\xi_{\alpha_1 \alpha_2 \ldots \alpha_p ; \gamma ; \bar{\delta}} - \xi_{\alpha_1 \alpha_2 \ldots \alpha_p ; \bar{\delta} ; \gamma} = - \xi_{\lambda \alpha_2 \ldots \alpha_p} R^{\lambda}{}_{\alpha_1 \gamma \bar{\delta}}$$

$$- \xi_{\alpha_1 \lambda \alpha_3 \ldots \alpha_p} R^{\lambda}{}_{\alpha_2 \gamma \bar{\delta}} - \cdots - \xi_{\alpha_1 \alpha_2 \ldots \alpha_{p-1} \lambda} R^{\lambda}{}_{\alpha_p \gamma \bar{\delta}}$$

we have, by contraction with $g^{\gamma \bar{\delta}}$

$$g^{\gamma \bar{\delta}} \xi_{\alpha_1 \alpha_2 \ldots \alpha_p ; \gamma ; \bar{\delta}} - \xi_{\lambda \alpha_2 \ldots \alpha_p} R^{\lambda}{}_{\alpha_1} - \xi_{\alpha_1 \lambda \alpha_3 \ldots \alpha_p} R^{\lambda}{}_{\alpha_2}$$

$$- \cdots - \xi_{\alpha_1 \alpha_2 \ldots \alpha_{p-1} \lambda} R^{\lambda}{}_{\alpha_p} = 0$$

This equation shows that, the tensor

$$\eta_{i_1 i_2 \dots i_p} = \delta_{i_1}^{\alpha_1} \delta_{i_2}^{\alpha_2} \dots \delta_{i_p}^{\alpha_p} \xi_{\alpha_1 \alpha_2 \dots \alpha_p}$$

satisfies

$$g^{jk} \eta_{i_1 i_2 \dots i_p ; j ; k} - \sum_{s=1}^{p} \eta_{i_1 \dots i_{s-1} a i_{s+1} \dots i_p} R^a{}_{i_s}$$

$$- \sum_{s<t}^{1 \dots p} \xi_{i_1 \dots i_{s-1} a i_{s+1} \dots i_{t-1} b i_{t+1} \dots i_p} R^{ab}{}_{i_s i_t} = 0$$

and is harmonic.

THEOREM 8.16. In a compact Kaehler manifold, if the components of an anti-symmetric covariant pure tensor of type zero are complex analytic functions of coordinates, then it is harmonic. Also, if the components of an anti-symmetric covariant tensor of order p and of type p are complex conjugate of analytic functions of coordinates, then it is harmonic. (Eckmann and Guggenheimer [3]).

From all these, we conclude

THEOREM 8.17. In a compact Kaehler manifold, if the components

$$\xi_{\alpha_1 \alpha_2 \dots \alpha_p}$$

of an anti-symmetric tensor field

$$\xi_{i_1 i_2 \dots i_p}$$

of the form

$$\xi_{i_1 i_2 \dots i_p} = (\xi_{\alpha_1 \alpha_2 \dots \alpha_p}, 0, 0, \dots, 0, 0, \xi_{\bar{\alpha}_1 \bar{\alpha}_2 \dots \bar{\alpha}_p})$$

are complex analytic functions of coordinates, and the components

$$\xi_{\bar{\alpha}_1\bar{\alpha}_2\cdots\bar{\alpha}_p}$$

of this tensor are complex conjugate of analytic functions of coordinates, then the tensor $\xi_{i_1 i_2 \cdots i_p}$ is harmonic.

Thus, an anti-symmetric tensor field of the form

$$\xi_{i_1 i_2 \cdots i_p} = (\xi_{\alpha_1\alpha_2\cdots\alpha_p},\ 0,\ 0,\ \ldots,\ 0,\ 0,\ \xi_{\bar{\alpha}_1\bar{\alpha}_2\cdots\bar{\alpha}_p})$$

is harmonic if and only if the components

$$\xi_{\alpha_1\alpha_2\cdots\alpha_p}$$

are complex analytic functions of coordinates and the components

$$\xi_{\bar{\alpha}_1\bar{\alpha}_2\cdots\bar{\alpha}_p}$$

are complex analytic functions of conjugate coordinates.

9. KILLING VECTOR FIELDS

In a compact Kaehler manifold, it is again true, as in the real case, that a vector ξ^i is a Killing vector if and only if

(8.74) $$g^{jk}\xi^i{}_{;j;k} + R^i{}_l\xi^l = 0$$

and

(8.75) $$\xi^i{}_{;i} = 0$$

Now, for a vector field $\xi^i = (\xi^\alpha, \xi^{\bar{\alpha}})$, not necessarily self-adjoint, relations (8.74) split into

(8.76) $$g^{jk}\xi^\alpha{}_{;j;k} + R^\alpha{}_\beta\xi^\beta = 0$$

(8.77) $$g^{jk}\xi^{\bar{\alpha}}{}_{;j;k} + R^{\bar{\alpha}}{}_{\bar{\beta}}\xi^{\bar{\beta}} = 0$$

and hence the following conclusion:

THEOREM 8.18. In a compact Kaehler manifold, if $(\xi^{\alpha}, \xi^{\bar{\alpha}})$ is a Killing vector, then its adjoint $(\overline{\xi^{\bar{\alpha}}}, \overline{\xi^{\alpha}})$ is likewise a Killing vector.

Now, the secondary condition (8.75) is implied by the stronger conditions

$$\xi^{\alpha}{}_{;\alpha} = \xi^{\bar{\alpha}}{}_{;\bar{\alpha}} = 0$$

and this condition is very much stronger, as the following theorem will show.

THEOREM 8.19. If on a compact Kaehler manifold, for a vector $(\xi^{\alpha}, \xi^{\bar{\alpha}})$ we have

$$\xi^{\alpha}{}_{;\alpha} = 0 = \xi^{\bar{\alpha}}{}_{;\bar{\alpha}} \tag{8.78}$$

then

(i) the vector is a Killing vector if and only if the vectors

$$\eta^{i} = (\xi^{\alpha}, 0) \qquad \zeta^{i} = (0, \xi^{\bar{\alpha}}) \tag{8.79}$$

are Killing vectors,

(ii) if the latter are Killing vectors, then the components $\xi_{\alpha} = g_{\alpha\bar{\beta}}\xi^{\beta}$ are analytic in (z^{α}), and the components $\xi_{\bar{\alpha}} = g_{\bar{\alpha}\beta}\xi^{\beta}$ are analytic in $(\bar{z}^{\alpha})$,

(iii) if the vector is a Killing vector, then it is a parallel vector field,

(iv) the vector is a Killing vector if and only if the contravariant components ξ^{α} are analytic in the (z^{α}), and $\xi^{\bar{\alpha}}$ are analytic in the $(\bar{z}^{\alpha})$.

PROOF. Ad (i), if ξ^{i} is a Killing vector, then it satisfies (8.74) and (8.75), but under the additional assumptions (8.78) this makes the vectors (8.79) Killing vectors, and conversely.

Ad (ii), the vector η^{i} is a Killing vector, if and only if

$$\eta_{i;j} + \eta_{j;i} = 0$$

But, this implies

$$\xi_{\bar{\alpha};\beta} = 0$$

and this is equivalent with $\xi_{\bar{\alpha}}$ being independent of the (z^{α}), and similarly for ξ_{α}.

Ad (iii), by (ii) and Theorem 8.13, η^i and ζ^i are harmonic vectors, and so is $\xi^i = \eta^i + \zeta^i$. Hence

$$\xi_{i;j} + \xi_{j;i} = 0$$

in addition to $\xi_{i;j} - \xi_{j;i} = 0$, and therefore $\xi_{i;j} = 0$ and thus ξ^i is a parallel field.

Ad (iv), for a parallel vector field, we have

$$\xi^{i}{}_{;j} = 0$$

and

$$\xi^{\alpha}{}_{;\bar{\beta}} = 0 \qquad \xi^{\bar{\alpha}}{}_{;\beta} = 0$$

and thus ξ^{α} and $\xi^{\bar{\alpha}}$ are analytic. The converse is easy to prove.

10. KILLING TENSOR FIELDS

On a compact Kaehler manifold, an anti-symmetric tensor

$$\xi_{i_1 i_2 \cdots i_p}$$

is a Killing tensor if and only if

$$\text{(8.80)} \quad g^{jk}\xi_{i_1 i_2 \cdots i_p;j;k} + \frac{1}{p}\sum_{s=1}^{p}\xi_{i_1 i_{s-1} a i_{s+1} \cdots i_p} R^{a}{}_{i_s} + \frac{1}{p}\sum_{s<t}^{1\cdots p}\xi_{i_1 \cdots i_{s-1} a i_{s+1} \cdots i_{t-1} b i_{t+1} \cdots i_p} R^{ab}{}_{i_s i_t} = 0$$

and

$$\text{(8.81)} \quad \xi^{i}{}_{i_2 \cdots i_p;i} = 0$$

This implies first of all

THEOREM 8.20. The adjoint

$$C(\xi_{i_1 i_2 \ldots i_p})$$

of a Killing tensor $\xi_{i_1 i_2 \ldots i_p}$ is again a Killing tensor.

We now replace (8.81) by the stronger conditions:

$$\xi^{\alpha}{}_{i_2 \ldots i_p;\alpha} = \xi^{\bar{\alpha}}{}_{i_2 \ldots i_p;\bar{\alpha}} = 0 \tag{8.82}$$

and if we introduce the pure tensor of type h

$$p!\, \delta^{\alpha_1}_{[i_1} \cdots \delta^{\alpha_{p-h}}_{i_{p-h}} \delta^{\bar{\beta}_{p-h+1}}_{i_{p-h+1}} \cdots \delta^{\bar{\beta}_p}_{i_p]} \xi_{\alpha_1 \ldots \alpha_{p-h} \bar{\beta}_{p-h+1} \ldots \bar{\beta}_p} \tag{8.83}$$

then we first of all have

THEOREM 8.21. In a compact Kaehler manifold, if an anti-symmetric tensor field

$$\xi_{i_1 i_2 \ldots i_p}$$

is a Killing one and satisfies (8.82), then every pure tensor (8.83) of type h (h = 1, 2, ..., p) derived from $\xi_{i_1 i_2 \ldots i_p}$ is also a Killing tensor.

Similarly, as analogue to Theorem 8.17, we obtain as follows:

THEOREM 8.22. If for a tensor

$$(\xi_{\alpha_1 \alpha_2 \ldots \alpha_p}, 0, 0, \ldots, 0, 0, \xi_{\bar{\alpha}_1 \bar{\alpha}_2 \ldots \bar{\alpha}_p})$$

we have

$$g^{\alpha\bar{\beta}} \xi_{\alpha\alpha_2 \ldots \alpha_p;\bar{\beta}} = g^{\bar{\alpha}\beta} \xi_{\bar{\alpha}\bar{\alpha}_2 \ldots \bar{\alpha}_p;\beta} = 0 \tag{8.84}$$

then

(i) the tensor is a Killing tensor if and only if the pure tensor of type zero and the pure tensor of type p are both Killing tensors,

(ii) if the latter are Killing tensors, then their covariant components are analytic in the variables in (z^α) and $(\bar{z}^\alpha)$ respectively,

(iii) if the tensor is a Killing tensor, it is a parallel tensor field:

$$\xi^{i_1 i_2 \cdots i_p}{}_{;j} = 0$$

(iv) if the tensor is a Killing tensor, then the first contravariant components of the tensor are analytic in the variables (z^α) and the last in $(\bar{z}^\alpha)$ respectively.

11. THE TENSOR h_{ij}

The Kaehler metric

(8.85) $$ds^2 = g_{ij}dz^i dz^j = 2g_{\alpha\bar{\beta}}dz^\alpha d\bar{z}^\beta \qquad (g_{ij} = g_{ji})$$

with

$$g_{\alpha\beta} = g_{\bar{\alpha}\bar{\beta}} = 0 \qquad\qquad g_{\alpha\bar{\beta}} = \overline{g_{\bar{\alpha}\beta}}$$

has the law of transformation

$$g'_{\alpha\bar{\beta}} = \frac{\partial z^\lambda}{\partial z'^\alpha}\frac{\partial \bar{z}^\mu}{\partial \bar{z}'^\beta} g_{\lambda\bar{\mu}}$$

and if we put

(8.86) $$h_{\alpha\bar{\beta}} = -\, ig_{\alpha\bar{\beta}} \qquad\qquad h_{\bar{\beta}\alpha} = +\, ig_{\bar{\beta}\alpha}$$

then

$$h'_{\alpha\bar{\beta}} = \frac{\partial z^\lambda}{\partial z'^\alpha}\frac{\partial \bar{z}^\mu}{\partial \bar{z}'^\beta} h_{\lambda\bar{\mu}}$$

$$h'_{\bar{\beta}\alpha} = \frac{\partial \bar{z}^\mu}{\partial \bar{z}'^\beta}\frac{\partial z^\lambda}{\partial z'^\alpha} h_{\bar{\mu}\lambda}$$

This equation shows that, if we define an anti-symmetric quantity h_{ij} by

$$(8.87) \qquad (h_{ij}) = \begin{pmatrix} 0 & -ig_{\alpha\bar{\beta}} \\ +ig_{\bar{\beta}\alpha} & 0 \end{pmatrix}$$

then h_{ij} is an anti-symmetric covariant tensor.

Since we have

$$(\overline{h_{ij}}) = \begin{pmatrix} 0 & +i\overline{g_{\alpha\bar{\beta}}} \\ -i\overline{g_{\bar{\beta}\alpha}} & 0 \end{pmatrix} = \begin{pmatrix} 0 & +ig_{\beta\bar{\alpha}} \\ -ig_{\bar{\alpha}\beta} & 0 \end{pmatrix} = (h_{\bar{i}\bar{j}})$$

this tensor is self-adjoint.

If we define

$$(8.88) \qquad h^i{}_j = g^{ia}h_{aj}$$

and

$$(8.89) \qquad h^{ij} = g^{ia}g^{jb}h_{ab}$$

then the actual values of $h^i{}_j$ are given by

$$(8.90) \qquad (h^i{}_j) = \begin{pmatrix} +i\delta^\alpha_\beta & 0 \\ 0 & -i\delta^{\bar{\alpha}}_{\bar{\beta}} \end{pmatrix}$$

and those of h^{ij} by

$$(8.91) \qquad (h^{ij}) = \begin{pmatrix} 0 & +ig^{\alpha\bar{\beta}} \\ -ig^{\bar{\beta}\alpha} & 0 \end{pmatrix}$$

From (8.90), we obtain

$$(8.92) \qquad h^i{}_a h^a{}_j = -\delta^i_j$$

which may be also written as

(8.93) $$g_{ab}h^a{}_i h^b{}_j = g_{ij}$$

We can see from (8.90) that

$$h^i{}_j\xi^j = (i\xi^\alpha, -i\xi^{\bar{\alpha}})$$

and on putting

(8.94) $$P^i{}_j = \frac{1}{2}(\delta^i_j + ih^i{}_j)$$

(8.95) $$Q^i{}_j = \frac{1}{2}(\delta^i_j - ih^i{}_j)$$

we have

(8.96) $$P^i{}_j\xi^j = (0, \xi^{\bar{\alpha}})$$

(8.97) $$Q^i{}_j\xi^j = (\xi^\alpha, 0)$$

Moreover, we can easily verify the following relations:

$$P^i{}_j + Q^i{}_j = \delta^i_j$$

(8.98) $$P^i{}_a P^a{}_j = P^i{}_j \qquad P^i{}_a Q^a{}_j = 0$$

$$Q^i{}_a P^a{}_j = 0 \qquad Q^i{}_a Q^a{}_j = Q^i{}_j$$

and

(8.99) $$P_{ij} = Q_{ji} \qquad Q_{ij} = P_{ji}$$

where

$$P_{ij} = g_{ia}P^a{}_j \qquad Q_{ij} = g_{ia}Q^a{}_j$$

On the other hand, using

$$g_{\alpha\bar{\beta};\gamma} = g_{\alpha\bar{\beta};\bar{\gamma}} = 0$$

we obtain

(8.100) $$h_{ij;k} = 0 \qquad h^i{}_{j;k} = 0 \qquad h^{ij}{}_{;k} = 0$$

and

(8.101) $$P^i{}_{j;k} = 0 \qquad Q^i{}_{j;k} = 0$$

Now, from $h^i{}_{j;k} = 0$ and the Ricci identity:

$$h^i{}_{j;k;l} - h^i{}_{j;l;k} = h^a{}_j R^i{}_{akl} - h^i{}_a R^a{}_{jkl}$$

we have

(8.102) $$h^i{}_a R^a{}_{jkl} = h^a{}_j R^i{}_{akl}$$

which is equivalent with

(8.103) $$R^i{}_{jkl} = - h^i{}_a h^b{}_j R^a{}_{bkl}$$

Also from (8.103), we get

(8.104) $$R^{ij}{}_{kl} = h^i{}_a h^j{}_b R^{ab}{}_{kl}$$

and

(8.105) $$R^{ij}{}_{kl} = R^{ij}{}_{ab} h^a{}_k h^b{}_l$$

and consequently

(8.106) $$R^{ij}{}_{ab} h^a{}_k h^b{}_l = h^i{}_a h^j{}_b R^{ab}{}_{kl}$$

On the other hand, multiplying (8.102) by g^{jk} and contracting, we obtain

$$h^i{}_a R^a{}_l = h^{ab} R^i{}_{abl}$$

$$= \frac{1}{2} h^{ab}(R^i{}_{abl} - R^i{}_{bal})$$

or

(8.107) $$- h^i{}_a R^a{}_l = \frac{1}{2} h^{ab} R^i{}_{lab}$$

and from equation (8.107), if written in the form

$$-h_{ia}R^a{}_l = \frac{1}{2}h^{ab}R_{ilab}$$

and from the anti-symmetry of R_{ilab} in i and l, we obtain

$$h_{ia}R^a{}_l + h_{la}R^a{}_i = 0$$

or

(8.108) $$h^i{}_aR^a{}_l = h^a{}_lR^i{}_a$$

or

(8.109) $$R^i{}_l = -h^i{}_ah^b{}_lR^a{}_b$$

Now, from (8.102), we find

(8.110) $$P^i{}_aR^a{}_{jkl} = P^a{}_jR^i{}_{akl}$$

and

(8.111) $$Q^i{}_aR^a{}_{jkl} = Q^a{}_jR^i{}_{akl}$$

and multiplying (8.110) by $Q^j{}_b$ and contracting, we find

$$P^i{}_aQ^j{}_bR^a{}_{jkl} = 0$$

or

(8.112) $$P^i{}_aP^j{}_bR^{ab}{}_{kl} = 0$$

Similarly we can prove

(8.113) $$Q^i{}_aQ^j{}_bR^{ab}{}_{kl} = 0$$

Next, multiplying (8.110) by $P^j{}_b$ and contracting, we find

$$P^i{}_aP^j{}_bR^a{}_{jkl} = P^a{}_bR^i{}_{akl}$$

or

$$P^i{}_a Q^j{}_b R^{ab}{}_{kl} = Q^j{}_a R^{ia}{}_{kl} \tag{8.114}$$

and similarly

$$Q^i{}_a P^j{}_b R^{ab}{}_{kl} = P^j{}_a R^{ia}{}_{kl} \tag{8.115}$$

From (8.114) and (8.115), we get

$$2P^{[i}{}_a Q^{j]}{}_b R^{ab}{}_{kl} = R^{ij}{}_{kl} \tag{8.116}$$

and hence

$$2R^{ij}{}_{ab} P^a{}_{[k} Q^b{}_{l]} = R^{ij}{}_{kl} \tag{8.117}$$

and consequently

$$2P^{[i}{}_a Q^{j]}{}_b R^{ab}{}_{kl} = 2R^{ij}{}_{ab} P^a{}_{[k} Q^b{}_{l]} \tag{8.118}$$

Also, from (8.108), we have

$$P^i{}_a R^a{}_l = P^a{}_l R^i{}_a \tag{8.119}$$

and

$$Q^i{}_a R^a{}_l = Q^a{}_l R^i{}_a \tag{8.120}$$

Now, if a vector ξ_i is harmonic, then we have

$$g^{jk} \xi_{i;j;k} - \xi_a R^a{}_i = 0$$

but, from this, we can obtain

$$g^{jk} (h^a{}_i \xi_a)_{;j;k} - (h^b{}_a \xi_b) R^a{}_i = 0$$

$$g^{jk} (P^a{}_i \xi_a)_{;j;k} - (P^b{}_a \xi_b) R^a{}_i = 0$$

$$g^{jk} (Q^a{}_i \xi_a)_{;j;k} - (Q^b{}_a \xi_b) R^a{}_i = 0$$

and these equations show that, if ξ_i is harmonic, then $h^a{}_i \xi_a$, $P^a{}_i \xi_a$,

$Q^a{}_i \xi_a$ are also harmonic.

Similarly we can prove that if

$$\xi_{i_1 i_2 \ldots i_p}$$

is harmonic, then

$$h^{a_1}{}_{i_1} \ldots h^{a_p}{}_{i_p} \xi_{a_1 \ldots a_p}$$

and

$$P^{a_1}{}_{[i_1} \ldots P^{a_h}{}_{i_h} Q^{a_{h+1}}{}_{i_{h+1}} \ldots Q^{a_p}{}_{i_p]} \xi_{a_1 \ldots a_p}$$

are also harmonic.

Now, an anti-symmetric tensor

$$\xi_{i_1 i_2 \ldots i_p}$$

for $p \geq 2$ shall be called an <u>effective</u> tensor if it satisfies

$$h^{ij} \xi_{iji_3 \ldots i_p} = 0 \tag{8.121}$$

and since we have

$$h^{ij} = \begin{pmatrix} 0 & +ig^{\alpha\bar{\beta}} \\ -ig^{\bar{\beta}\alpha} & 0 \end{pmatrix}$$

we can see that (8.121) is equivalent to

$$g^{\alpha\bar{\beta}} \xi_{\alpha\bar{\beta} i_3 \ldots i_p} = 0 \tag{8.122}$$

Regarding effective harmonic tensor in a compact Kaehler manifold, Hodge [1] has proved the following fundamental theorem which we will utilize.

THEOREM 8.23. In a compact Kaehler manifold of real dimension $2n$, we have

$$B_0 \leq B_2 \leq B_4 \leq \cdots \leq B_{2[n/2]}$$

$$B_1 \leq B_3 \leq B_5 \leq \cdots \leq B_{2[n/2]+1}$$

and the number $B_{p+2} - B_p$ $(p + 2 \leq n)$ is the number of linearly independent (with constant coefficients) self-adjoint effective harmonic tensors of order $p + 2$.

Thus, if there are no self-adjoint effective harmonic tensors, then

$$B_0 = B_2 = B_4 = \cdots = B_{2[n/2]}$$

$$B_1 = B_3 = B_5 = \cdots = B_{2[n/2]+1}$$

12. EFFECTIVE HARMONIC TENSORS IN FLAT MANIFOLDS

We are returning to an analysis of the quadratic form

$$(8.123) \quad F(\xi_{i_1 i_2 \cdots i_p}) = R_{ij}\xi^{ii_2 \cdots i_p}\xi^{j}{}_{i_2 \cdots i_p} + \frac{p-1}{2} R_{ijkl}\xi^{iji_3 \cdots i_p}\xi^{kl}{}_{i_3 \cdots i_p}$$

for an anti-symmetric tensor

$$\xi_{i_1 i_2 \cdots i_p}$$

and we are interested in estimating it, in Kaehler metric, for a tensor which is self-adjoint and effective. The application will be that whenever no such exists, then by Theorem 8.23, we must have

$$B_{p-2} = B_p$$

For

$$R_{\alpha\bar{\beta}\gamma\bar{\delta}} = \frac{k}{2}(g_{\alpha\bar{\beta}}g_{\gamma\bar{\delta}} + g_{\alpha\bar{\delta}}g_{\gamma\bar{\beta}}) \qquad (k > 0)$$

we have also

$$R_{\alpha\bar{\beta}} = \frac{n+1}{2} k g_{\alpha\bar{\beta}}$$

and therefore

$$R_{ij}\xi^{ii_2\cdots i_p}{}_\xi{}^j{}_{i_2\cdots i_p} = 2(R_{\alpha\bar\beta}\xi^{\alpha\gamma i_3\cdots i_p}{}_\xi{}^{\bar\beta}{}_{\gamma i_3\cdots i_p} + R_{\alpha\bar\beta}\xi^{\alpha\bar\gamma i_3\cdots i_p}{}_\xi{}^{\bar\beta}{}_{\bar\gamma i_3\cdots i_p})$$

$$(8.124) \qquad = (n+1)k(\xi^{\alpha\beta i_3\cdots i_p}\xi_{\alpha\beta i_3\cdots i_p} + \xi^{\alpha\bar\beta i_3\cdots i_p}\xi_{\alpha\bar\beta i_3\cdots i_p})$$

On the other hand, we have

$$R_{ijkl}\xi^{iji_3\cdots i_p}{}_\xi{}^{kl}{}_{i_3\cdots i_p} = 4R_{\alpha\bar\beta\gamma\bar\delta}\xi^{\alpha\bar\beta i_3\cdots i_p}{}_\xi{}^{\gamma\bar\delta}{}_{i_3\cdots i_p}$$

and consequently

$$(8.125) \qquad R_{ijkl}\xi^{iji_3\cdots i_p}{}_\xi{}^{kl}{}_{i_3\cdots i_p} = -2k\xi^{\alpha\bar\beta i_3\cdots i_p}\xi_{\alpha\bar\beta i_3\cdots i_p}$$

by virtue of

$$g_{\alpha\bar\beta}\xi^{\alpha\bar\beta i_3\cdots i_p} = 0$$

Altogether we have

$$F(\xi_{i_1 i_2\cdots i_p}) = (n+1)k\xi^{\alpha\beta i_3\cdots i_p}\xi_{\alpha\beta i_3\cdots i_p}$$

$$+ (n-p+2)k\xi^{\alpha\bar\beta i_3\cdots i_p}\xi_{\alpha\bar\beta i_3\cdots i_p}$$

which is positive for $0 \le p \le n$.

THEOREM 8.24. In a compact Kaehler manifold of positive constant holomorphic curvature $k > 0$, we have

$$B_{2l} = 1 \qquad B_{2l+1} = 0 \qquad (0 \le 2l\ ,\ 2l+1 \le n)$$

We will now envisage as formal analogues to the Weyl projective and conformal curvature tensors, the following tensors:

$$(8.126) \qquad G_{\alpha\bar\beta\gamma\bar\delta} = R_{\alpha\bar\beta\gamma\bar\delta} - \frac{1}{n+1}(g_{\alpha\bar\beta}R_{\gamma\bar\delta} + g_{\alpha\bar\delta}R_{\gamma\bar\beta})$$

$$(8.127)\quad H_{\alpha\bar\beta\gamma\bar\delta} = R_{\alpha\bar\beta\gamma\bar\delta} - \frac{1}{2(n+1)}(g_{\alpha\bar\beta}R_{\gamma\bar\delta} + g_{\alpha\bar\delta}R_{\gamma\bar\beta} + g_{\gamma\bar\delta}R_{\alpha\bar\beta} + g_{\gamma\bar\beta}R_{\alpha\bar\delta})$$

and

$$(8.128)\quad K_{\alpha\bar\beta\gamma\bar\delta} = R_{\alpha\bar\beta\gamma\bar\delta} - \frac{1}{n+2}(g_{\alpha\bar\beta}R_{\gamma\bar\delta} + g_{\alpha\bar\delta}R_{\gamma\bar\beta} + g_{\gamma\bar\delta}R_{\alpha\bar\beta} + g_{\gamma\bar\beta}R_{\alpha\bar\delta})$$
$$+ \frac{R}{2(n+1)(n+2)}(g_{\alpha\bar\beta}g_{\gamma\bar\delta} + g_{\alpha\bar\delta}g_{\gamma\bar\beta})$$

all of which were introduced in Bochner [6].

These tensors satisfy

$$(8.129)\qquad g^{\alpha\bar\delta}G_{\alpha\bar\beta\gamma\bar\delta} = 0 \qquad\qquad g^{\alpha\bar\delta}K_{\alpha\bar\beta\gamma\bar\delta} = 0$$

and since, for a manifold of constant holomorphic curvature, we have

$$R_{\alpha\bar\beta\gamma\bar\delta} = \frac{R}{2n(n+1)}(g_{\alpha\bar\beta}g_{\gamma\bar\delta} + g_{\alpha\bar\delta}g_{\gamma\bar\beta})$$

$$R_{\alpha\bar\beta} = \frac{R}{2n}g_{\alpha\bar\beta}$$

it is evident that, for such a manifold, we have

$$G_{\alpha\bar\beta\gamma\bar\delta} = 0 \qquad\qquad H_{\alpha\bar\beta\gamma\bar\delta} = 0$$

Conversely, if we assume that

$$G_{\alpha\bar\beta\gamma\bar\delta} = 0$$

then

$$(8.130)\qquad R_{\alpha\bar\beta\gamma\bar\delta} = \frac{1}{n+1}(g_{\alpha\bar\beta}R_{\gamma\bar\delta} + g_{\alpha\bar\delta}R_{\gamma\bar\beta})$$

and on substituting this into

$$R_{\alpha\bar\beta\gamma\bar\delta} = R_{\gamma\bar\beta\alpha\bar\delta}$$

we find

$$g_{\alpha\bar\beta}R_{\gamma\bar\delta} + g_{\alpha\bar\delta}R_{\gamma\bar\beta} = g_{\gamma\bar\beta}R_{\alpha\bar\delta} + g_{\gamma\bar\delta}R_{\alpha\bar\beta}$$

Multiplying this by $g^{\alpha\bar\beta}$ and contracting, we find

$$nR_{\gamma\bar{\delta}} + R_{\gamma\bar{\delta}} = R_{\gamma\bar{\delta}} + g_{\gamma\bar{\delta}} \frac{R}{2}$$

or

$$R_{\gamma\bar{\delta}} = \frac{R}{2n} g_{\gamma\bar{\delta}}$$

and, by (8.130), we obtain

$$R_{\alpha\bar{\beta}\gamma\bar{\delta}} = \frac{R}{2n(n+1)} (g_{\alpha\bar{\beta}}g_{\gamma\bar{\delta}} + g_{\alpha\bar{\delta}}g_{\gamma\bar{\beta}})$$

the conclusion then being that the manifold is of constant holomorphic curvature.

Similarly for

$$H_{\alpha\bar{\beta}\gamma\bar{\delta}} = 0$$

we obtain

(8.131) $$R_{\alpha\bar{\beta}\gamma\bar{\delta}} = \frac{1}{2(n+1)} (g_{\alpha\bar{\beta}}R_{\gamma\bar{\delta}} + g_{\alpha\bar{\delta}}R_{\gamma\bar{\beta}} + g_{\gamma\bar{\delta}}R_{\alpha\bar{\beta}} + g_{\gamma\bar{\beta}}R_{\alpha\bar{\delta}})$$

and by contraction with $g^{\alpha\bar{\delta}}$,

$$R_{\bar{\beta}\gamma} = \frac{R}{2n} g_{\bar{\beta}\gamma}$$

and then from (8.131),

$$R_{\alpha\bar{\beta}\gamma\bar{\delta}} = \frac{R}{2n(n+1)} (g_{\alpha\bar{\beta}}g_{\gamma\bar{\delta}} + g_{\alpha\bar{\delta}}g_{\gamma\bar{\beta}})$$

and thus the manifold is again of constant holomorphic curvature.

For

(8.132) $$K_{\alpha\bar{\beta}\gamma\bar{\delta}} = 0$$

that is,

$$R_{\alpha\bar{\beta}\gamma\bar{\delta}} = \frac{1}{n+2} (g_{\alpha\bar{\beta}}R_{\gamma\bar{\delta}} + g_{\alpha\bar{\delta}}R_{\gamma\bar{\beta}} + g_{\gamma\bar{\delta}}R_{\alpha\bar{\beta}} + g_{\gamma\bar{\beta}}R_{\alpha\bar{\delta}})$$

$$- \frac{R}{2(n+1)(n+2)} (g_{\alpha\bar{\beta}}g_{\gamma\bar{\delta}} + g_{\alpha\bar{\delta}}g_{\gamma\bar{\beta}})$$

we do not claim this conclusion, but the effect of (8.132) on Betti numbers will be the same as of constant holomorphic curvature.

In fact, for an effective tensor, we have

$$R_{ij}\xi^{ii_2\ldots i_p}\xi^{j}{}_{i_2\ldots i_p} = 2(R_{\alpha\bar\beta}\xi^{\alpha\gamma i_3\ldots i_p}\xi^{\bar\beta}{}_{\gamma i_3\ldots i_p} + R_{\alpha\bar\beta}\xi^{\alpha\bar\gamma i_3\ldots i_p}\xi^{\bar\beta}{}_{\bar\gamma i_3\ldots i_p})$$

and

$$R_{ijkl}\xi^{iji_3\ldots i_p}\xi^{kl}{}_{i_3\ldots i_p} = 4R_{\alpha\bar\beta\gamma\bar\delta}\xi^{\alpha\bar\beta i_3\ldots i_p}\xi^{\gamma\bar\delta}{}_{i_3\ldots i_p}$$

$$= -\frac{8}{n+2}R_{\alpha\bar\beta}\xi^{\alpha\bar\gamma i_3\ldots i_p}\xi^{\bar\beta}{}_{\bar\gamma i_3\ldots i_p}$$

$$+\frac{2R}{(n+1)(n+2)}\xi^{\alpha\bar\beta i_3\ldots i_p}\xi_{\alpha\bar\beta i_3\ldots i_p}$$

and consequently

$$(8.133)\quad F(\xi_{i_1 i_2\ldots i_p}) = 2R_{\alpha\bar\beta}\xi^{\alpha\gamma i_3\ldots i_p}\xi^{\bar\beta}{}_{\gamma i_3\ldots i_p}$$

$$+ 2\left[1-\frac{2(p-1)}{n+2}\right]R_{\alpha\bar\beta}\xi^{\alpha\bar\gamma i_3\ldots i_p}\xi^{\bar\beta}{}_{\bar\gamma i_3\ldots i_p}$$

$$+\frac{p-1}{(n+1)(n+2)}R\xi^{\alpha\bar\beta i_3\ldots i_p}\xi_{\alpha\bar\beta i_3\ldots i_p}$$

and hence the following conclusion:

THEOREM 8.25. In a compact Kaehler manifold in which $K_{\alpha\bar\beta\gamma\bar\delta} = 0$ and $R_{\alpha\bar\beta}\xi^{\alpha}\xi^{\bar\beta}$ is positive definite, we have

$$F(\xi_{i_1 i_2\ldots i_p}) \geq 0$$

for $1 \leq p \leq n/2 + 2$, and therefore

$$B_{2l} = 1 \qquad B_{2l+1} = 0 \qquad (0 \leq 2l\ ,\ 2l+1 \leq \tfrac{n}{2} + 2)$$

(Bochner [6]).

In order to obtain a result for all p, we introduce here a tensor

(8.134) $$S_{\alpha\bar{\beta}} = R_{\alpha\bar{\beta}} - \frac{R}{2n} g_{\alpha\bar{\beta}}$$

and the quantity

(8.135) $$S = \sup_{\xi} \frac{|S_{\alpha\bar{\beta}}\xi^{\alpha}\xi^{\bar{\beta}}|}{\xi^{\alpha}\xi_{\alpha}}$$

which measures deviation from being an Einstein manifold.

Substituting

$$R_{\alpha\bar{\beta}} = S_{\alpha\bar{\beta}} + \frac{R}{2n} g_{\alpha\bar{\beta}}$$

in (8.133), we get

$$F(\xi_{i_1 i_2 \cdots i_p}) = 2R_{\alpha\bar{\beta}}\xi^{\alpha\gamma i_3 \cdots i_p}\xi^{\bar{\beta}}{}_{\gamma i_3 \cdots i_p} + \frac{n-p+2}{n(n+1)} R\xi^{\alpha\bar{\beta} i_3 \cdots i_p}\xi_{\alpha\bar{\beta} i_3 \cdots i_p}$$

$$+ \frac{2(n-2p+4)}{n+2} S_{\alpha\bar{\beta}}\xi^{\alpha\bar{\gamma} i_3 \cdots i_p}\xi^{\bar{\beta}}{}_{\bar{\gamma} i_3 \cdots i_p}$$

But, for

$$\frac{n}{2} + 3 \leq p \leq n \qquad (6 \leq n)$$

we have

$$2 \leq n - p + 2 \leq \frac{n}{2} - 1 \qquad - 2(n - 4) \leq 2(n - 2p + 4) \leq - 4$$

and consequently

(8.136) $$F(\xi_{i_1 i_2 \cdots i_p}) \geq 2R_{\alpha\bar{\beta}}\xi^{\alpha\gamma i_3 \cdots i_p}\xi^{\bar{\beta}}{}_{\gamma i_3 \cdots i_p}$$

$$+ 2\left[\frac{1}{n(n+1)} R - \frac{n-4}{n+2} S\right]\xi^{\alpha\bar{\beta} i_3 \cdots i_p}\xi_{\alpha\bar{\beta} i_3 \cdots i_p}$$

Hence the following theorem:

THEOREM 8.26. If in a compact Kaehler manifold with $R_{\alpha\bar{\beta}}\xi^{\alpha}\xi^{\bar{\beta}} \geq 0$, we have

$$K_{\alpha\bar{\beta}\gamma\bar{\delta}} = 0$$

and

$$\frac{1}{n(n+1)} R > \frac{n-4}{n+2} S$$

then

$$B_{2l} = 1 \qquad B_{2l+1} = 0$$

for all dimensions.

13. DEVIATION FROM FLATNESS

We introduce the quantity

$$G = \sup_{\xi} \frac{|G_{\alpha\bar{\beta}\gamma\bar{\delta}}\xi^{\alpha\bar{\beta}}\xi^{\gamma\bar{\delta}}|}{\xi^{\alpha\bar{\beta}}\xi_{\alpha\bar{\beta}}} \tag{8.137}$$

and

$$H = \sup_{\xi} \frac{|H_{\alpha\bar{\beta}\gamma\bar{\delta}}\xi^{\alpha\bar{\beta}}\xi^{\gamma\bar{\delta}}|}{\xi^{\alpha\bar{\beta}}\xi_{\alpha\bar{\beta}}} \tag{8.138}$$

and for $R_{\alpha\bar{\beta}}\xi^{\alpha}\xi^{\bar{\beta}} \geq 0$, we put

$$L = \inf_{\xi} \frac{R_{\alpha\bar{\beta}}\xi^{\alpha}\xi^{\bar{\beta}}}{\xi^{\alpha}\xi_{\alpha}} \tag{8.139}$$

For an effective tensor we then have

$$\begin{aligned}
F(\xi_{i_1 i_2 \cdots i_p}) &= 2\Big[R_{\alpha\bar{\beta}}\xi^{\alpha i_2\cdots i_p}\xi^{\bar{\beta}}{}_{i_2\cdots i_p} + (p-1)R_{\alpha\bar{\beta}\gamma\bar{\delta}}\xi^{\alpha\bar{\beta}i_3\cdots i_p}\xi^{\gamma\bar{\delta}}{}_{i_3\cdots i_p}\Big] \\
&= 2\Big[R_{\alpha\bar{\beta}}\xi^{\alpha\gamma i_3\cdots i_p}\xi^{\bar{\beta}}{}_{\gamma i_3\cdots i_p} + \left(1 - \frac{p-1}{n+1}\right)R_{\alpha\bar{\beta}}\xi^{\alpha\bar{\gamma} i_3\cdots i_p}\xi^{\bar{\beta}}{}_{\bar{\gamma} i_3\cdots i_p} \\
&\qquad + (p-1)G_{\alpha\bar{\beta}\gamma\bar{\delta}}\xi^{\alpha\bar{\beta}i_3\cdots i_p}\xi^{\gamma\bar{\delta}}{}_{i_3\cdots i_p}\Big] \\
&\geq 2\Big[R_{\alpha\bar{\beta}}\xi^{\alpha\gamma i_3\cdots i_p}\xi^{\bar{\beta}}{}_{\gamma i_3\cdots i_p} \\
&\qquad + \left\{\left(1 - \frac{p-1}{n+1}\right)L - (p-1)G\right\}\xi^{\alpha\bar{\beta}i_3\cdots i_p}\xi_{\alpha\bar{\beta}i_3\cdots i_p}\Big]
\end{aligned}$$

and from this, and a similar estimate involving H, we obtain

THEOREM 8.27. In a compact Kaehler manifold, if $R_{\alpha\bar{\beta}}\xi^{\alpha}\xi^{\bar{\beta}}$ is positive definite, and

$$(1 - \frac{p-1}{n+1})L > (p - 1)G \quad \text{or} \quad (1 - \frac{p-1}{n+1})L > (p - 1)H \tag{8.140}$$

for

$$1 \leq p \leq n \tag{8.141}$$

then there exists no effective harmonic tensor of order p.

Also, if (8.140) holds for all (8.141), then we have

$$B_{2l} = 1 \qquad B_{2l+1} = 0 \qquad (0 \leq 2l \ , \ 2l+1 \leq n)$$

Finally we introduce the quantity

$$K = \sup_{\xi} \frac{|K_{\alpha\bar{\beta}\gamma\bar{\delta}}\xi^{\alpha\bar{\beta}}\xi^{\gamma\bar{\delta}}|}{\xi^{\alpha\bar{\beta}}\xi_{\alpha\bar{\beta}}}$$

and we claim as follows:

THEOREM 8.28. In a compact Kaehler manifold with $R_{\alpha\bar{\beta}}\xi^{\alpha}\xi^{\bar{\beta}} \geq 0$, if

$$(1 - \frac{2(p-1)}{n+2})L + \frac{(p-1)}{2(n+1)(n+2)} R > (p - 1)K$$

holds for

$$1 \leq p \leq \frac{n}{2} + 2$$

and if also

$$\frac{1}{n(n+1)} R - \frac{n-4}{n+2} S > (n - 1)K$$

then we have

$$B_{2l} = 1 \qquad B_{2l+1} = 0$$

for

$$0 \leq 2l \qquad 2l + 1 \leq n$$

PROOF. We will distinguish between the cases

(8.142) $$0 \leq 2l \qquad 2l + 1 \leq \frac{n}{2} + 2$$

and

(8.143) $$\frac{n}{2} + 3 \leq 2l \qquad 2l + 1 \leq n$$

and we start from the joint formula, for an effective tensor

$$F\{\xi_{i_1 i_2 \cdots i_p}\} = 2\left[R_{\alpha\bar{\beta}}\xi^{\alpha i_2 \cdots i_p}\xi^{\bar{\beta}}{}_{i_2 \cdots i_p} + (p-1)R_{\alpha\bar{\beta}\gamma\bar{\delta}}\xi^{\alpha\bar{\beta} i_3 \cdots i_p}\xi^{\gamma\bar{\delta}}{}_{i_3 \cdots i_p}\right]$$

(8.144) $$= 2\Big[R_{\alpha\bar{\beta}}\xi^{\alpha\gamma i_3 \cdots i_p}\xi^{\bar{\beta}}{}_{\gamma i_3 \cdots i_p}$$

$$+ \left(1 - \frac{2(p-1)}{n+2}\right)R_{\alpha\bar{\beta}}\xi^{\alpha\bar{\gamma} i_3 \cdots i_p}\xi^{\bar{\beta}}{}_{\bar{\gamma} i_3 \cdots i_p}$$

$$+ \frac{(p-1)R}{2(n+1)(n+2)}\xi^{\alpha\bar{\beta} i_3 \cdots i_p}\xi_{\alpha\bar{\beta} i_3 \cdots i_p}$$

$$+ (p-1)K_{\alpha\bar{\beta}\gamma\bar{\delta}}\xi^{\alpha\bar{\beta} i_3 \cdots i_p}\xi^{\gamma\bar{\delta}}{}_{i_3 \cdots i_p}\Big]$$

Thus, for

$$1 - \frac{2(p-1)}{n+2} \geq 0 \quad \text{or} \quad 1 \leq p \leq \frac{n}{2} + 2$$

we have

$$F\{\xi_{i_1 i_2 \cdots i_p}\} \geq 2\Big[R_{\alpha\bar{\beta}}\xi^{\alpha\gamma i_3 \cdots i_p}\xi^{\bar{\beta}}{}_{\gamma i_3 \cdots i_p}$$

$$+ \left\{\left(1 - \frac{2(p-1)}{n+2}\right)L + \frac{(p-1)}{2(n+1)(n+2)}R - (p-1)K\right\}\xi^{\alpha\bar{\beta} i_3 \cdots i_p}\xi_{\alpha\bar{\beta} i_3 \cdots i_p}\Big]$$

and this proves our assertion for (8.142).

Next, if in (8.144), we substitute

$$R_{\alpha\bar{\beta}} = S_{\alpha\bar{\beta}} + \frac{R}{2n} g_{\alpha\bar{\beta}}$$

we find

$$F(\xi_{i_1 i_2 \cdots i_p}) = 2\Big[R_{\alpha\bar{\beta}} \xi^{\alpha\gamma i_3 \cdots i_p} \xi^{\bar{\beta}}{}_{\gamma i_3 \cdots i_p} + \frac{n-p+2}{2n(n+1)} R \xi^{\alpha\bar{\beta} i_3 \cdots i_p} \xi_{\alpha\bar{\beta} i_3 \cdots i_p}$$

$$(8.145) \qquad + \frac{n-2p+4}{n+2} S_{\alpha\bar{\beta}} \xi^{\alpha\bar{\gamma} i_3 \cdots i_p} \xi^{\bar{\beta}}{}_{\bar{\gamma} i_3 \cdots i_p}$$

$$+ (p-1) K_{\alpha\bar{\beta}\gamma\bar{\delta}} \xi^{\alpha\bar{\beta} i_3 \cdots i_p} \xi^{\gamma\bar{\delta}}{}_{i_3 \cdots i_p} \Big]$$

and for $n/2 + 3 \leq p \leq n$, this is

$$(8.146) \quad F(\xi_{i_1 i_2 \cdots i_p}) \geq 2\Big[R_{\alpha\bar{\beta}} \xi^{\alpha\gamma i_3 \cdots i_p} \xi^{\bar{\beta}}{}_{\gamma i_3 \cdots i_p}$$

$$+ \Big(\frac{1}{n(n+1)} R - \frac{n-4}{n+2} S - (n-1) K\Big) \xi^{\alpha\bar{\beta} i_3 \cdots i_p} \xi_{\alpha\bar{\beta} i_3 \cdots i_p} \Big]$$

and hence the conclusion in case (8.143).

CHAPTER IX

SUPPLEMENTS

S. Bochner

1. SYMMETRIC MANIFOLDS

On a Riemannian manifold we take an arbitrary tensor but we immediately denote it by R_{ijkl}, because it will very soon be the curvature tensor itself. For the square length

$$\phi = R^{ijkl}R_{ijkl}$$

we form the Laplacean $\Delta\phi$, and the quantity $\frac{1}{2}\Delta\phi$ is then the sum of

$$g^{rs}R^{ijkl}{}_{;r}R_{ijkl;s} \tag{9.1}$$

and of

$$R^{ijkl}g^{rs}R_{ijkl;r;s} \tag{9.2}$$

Now, the value of (9.1) is ≥ 0, and it is $= 0$ if and only if

$$R_{ijkl;r} = 0 \tag{9.3}$$

and if, on a compact manifold, the quantity (9.2) is ≥ 0, then by Theorem 2.3, both must be zero. However (9.2) is zero in particular if

$$R_{ijkl;r;s} = 0 \tag{9.4}$$

which is a weaker condition than (9.3), and if we apply this reasoning to an arbitrary tensor we first of all obtain the following useful lemma (Bochner [7]):

THEOREM 9.1. On a compact Riemannian manifold, if for an arbitrary tensor (or scalar), the r-th successive covariant derivative vanishes,

$$\xi^{i_1 i_2 \cdots i_p}{}_{j_1 j_2 \cdots j_q ; k_1 ; k_2 ; \cdots ; k_r} = 0$$

then the first derivative already vanishes, that is,

$$\xi^{i_1 i_2 \cdots i_p}{}_{j_1 j_2 \cdots j_q ; k} = 0$$

Assume now that R_{ijkl} is the curvature tensor. In this case conditions (9.3) express the fact that the manifold is a symmetric space so-called, and by contracting (9.3) with g^{il} we obtain in particular also

(9.5) $$R_{jk;l} = 0$$

where R_{jk} is the Ricci tensor.

If now we introduce the Bianchi identity:

$$R_{ijkl;r} = - R_{ijlr;k} - R_{ijrk;l}$$

and form its covariant derivative and then again apply this identity, and in-between apply the Ricci identity judiciously, then it can be shown that the term (9.2) is the sum of the term

(9.6) $$4R^{ijkl}R_{jk;i;l}$$

and of the term

(9.7) $$H(R) = 2R^{ijkl}H_{ijk}{}^{s}{}_{ls}$$

where

(9.8) $$- H_{ijklrs} \equiv R_{ijkl;r;s} - R_{ijkl;s;r}$$

$$= - R_{ajkl}R^{a}{}_{irs} - R_{iakl}R^{a}{}_{jrs} - R_{ijal}R^{a}{}_{krs} - R_{ijka}R^{a}{}_{lrs}$$

The outcome of this decomposition is that the form $H(R)$ no longer involves derivatives of the curvature tensor and that (9.6) involves those of the "reduced" Ricci tensor only. Also, since for a symmetric space, (9.2) must be zero and (9.5) holds, both terms (9.6) and $H(R)$ must then be zero, and thus it is of interest to state the following converse conclusion now arising.

THEOREM 9.2. On a compact Riemannian manifold, if we have $R_{ij;r} = 0$, and if we have $H(R) \geq 0$, then we must have $R_{ijkl;r} = 0$, (symmetric space).

Somewhat more general than symmetric spaces are so-called recurrent spaces,

$$R_{ijkl;r} = R_{ijkl}\lambda_r$$

which were introduced by H. S. Ruse [1] and A. G. Walker [1], and even the more general spaces

$$R_{ijkl;r;s} = R_{ijkl}a_{rs}$$

might be envisaged; and for such spaces very close analogues to Theorem 9.2 can be formulated for which we refer the reader to Lichnérowicz [9].

2. CONVEXITY

If a Riemannian manifold V_n is isometrically immersed in the Euclidean E_{n+r}, then there exist on V_n, r symmetric tensors $b_{ij}^{(\rho)}$, $\rho = 1, 2, \ldots, r$, such that for the curvature tensor on V_n we have the representation

$$R_{ijkl} = \sum_{\rho=1}^{r} (b_{jk}^{(\rho)}b_{il}^{(\rho)} - b_{jl}^{(\rho)}b_{ik}^{(\rho)}) \tag{9.9}$$

Now, we assume, more generally, that for our V_n, the curvature tensor has such a representation in the neighborhood of every point, with tensors $b_{ij}^{(\rho)}$ defined in each neighborhood only; and, for the moment, we will call V_n <u>intrinsically semi-convex</u> if all $b_{ij}^{(\rho)}$ are positive semi-definite, and we call it <u>intrinsically convex</u> if at least one tensor is positive definite. Also, in the semi-convex case, we may more generally assume, that in every neighborhood, R_{ijkl} is only a uniform limit of expressions (9.9) in which each $b_{ij}^{(\rho)}$ is positive semi-definite.

Now by examining the expression

$$F(\xi_{ij}) = R_{ij}\xi^{ia}\xi^{j}_{a} + \frac{1}{2} R_{ijkl}\xi^{ij}\xi^{kl}$$

for a curvature tensor of our form (9.9), we obtain, as a corollary to Theorem 3.4, the following conclusion:

THEOREM 9.3. If a compact orientable Riemannian

manifold V_n is intrinsically convex, then the first two Betti numbers are zero, $B_1 = B_2 = 0$, and if it is intrinsically semi-convex, then harmonic vectors ξ_i and tensor ξ_{ij} must have covariant derivative zero.

For $r = 1$, in (9.9), one can even make a statement about all Betti numbers as follows:

THEOREM 9.4. If a compact orientable Riemannian manifold V_n is convexly isometrically imbedded in a Euclidean E_{n+1}, and if the smallest and largest principal curvatures differ at most by a factor 2, then all Betti numbers are zero.

This theorem was a stated in Lichnérowicz [9], and it also is obtainable from Theorem 5.1 of the present work.

3. MINIMAL VARIETIES

If we assume that the individual tensors $b_{ij}^{(\rho)}$ in (9.9) are all negative definite, we can draw no immediate conclusion, at least for real varieties which we are now considering; but a conclusion can be obtained if we assume that we have

$$g^{il}b_{il}^{(\rho)} = 0 \qquad \rho = 1, 2, \ldots, r \tag{9.10}$$

because in this case we have

$$R_{jk}\xi^j\xi^k = \sum_\rho g^{il}\eta_i^{(\rho)}\eta_l^{(\rho)}$$

where

$$\eta_i^{(\rho)} = b_{ik}^{(\rho)}\xi^k$$

Now, we will for the moment say that V_n is <u>intrinsically minimal</u>, if in the nieghborhood of every point we have a representation (9.9) with (9.10) holding, and will say that V_n is <u>intrinsically strictly minimal</u> if at every point at least one $b_{ik}^{(\rho)}$ has non vanishing determinant. We can now state

THEOREM 9.5. If a compact Riemannian V_n is intrinsically strictly minimal then there exists on it no

Killing vector and hence no one-parameter group of motions. And if the manifold is minimal, then the only groups of motions are translations and the orbits must be geodesics not only on the V_n but also on the locally envelopping Euclidean E_{n+r}. (Bochner [2]).

4. COMPLEX IMBEDDING

On a complex neighborhood $(z^1, z^2, \ldots, z^n)$ we take r covariant analytic vector fields

$$\zeta_\alpha^{(\rho)}(z) \qquad \rho = 1, 2, \ldots, r \tag{9.11}$$

$r \geq n$ and we make the following two assumptions. For each vector we have the curl identity:

$$\frac{\partial \zeta_\alpha^{(\rho)}}{\partial z^\beta} = \frac{\partial \zeta_\beta^{(\rho)}}{\partial z^\alpha} \qquad \alpha, \beta = 1, 2, \ldots, n \; ; \; \rho = 1, 2, \ldots, r \tag{9.12}$$

and, secondly, the matrix

$$\{\zeta_\alpha^{(\rho)}\} \tag{9.13}$$

or r rows and n columns has everywhere its maximal rank n.

If now we put

$$g_{\alpha\bar\beta} = \sum_{\rho=1}^{r} \zeta_\alpha^{(\rho)} \overline{\zeta_\beta^{(\rho)}} \tag{9.14}$$

then due to the second assumption this matrix is positive definite, and due to the assumption (9.12) which is a decisive one it has the Kaehler property; and the genesis of the resulting metric

$$ds^2 = g_{\alpha\bar\beta} dz^\alpha d\bar z^\beta \tag{9.15}$$

is as follows. Assumption (9.12) in the neighborhood of a point is the necessary and sufficient condition for the existence of analytic functions

$$w^\rho = f^\rho(z^1 z^2, \ldots, z^n) \tag{9.16}$$

for which

$$\zeta_\alpha^{(\rho)}. = \frac{\partial f^\rho}{\partial z^\alpha} \tag{9.17}$$

and if we interpret (9.16) as a mapping of our neighborhood into the manifold of the variables w^{ρ} , then by the second assumption the mapping is non singular and the metric (9.15) is induced by the flat Kaehler metric

$$(9.18) \qquad dw^1 d\bar{w}^1 + dw^2 d\bar{w}^2 + \ldots + dw^r d\bar{w}^r$$

of the envelopping manifold.

Now, the salient feature of this metric (9.15) is the fact that its Ricci tensor has the form

$$(9.19) \qquad - R_{\alpha\bar{\beta}} = g^{\gamma\bar{\delta}} \sum_{\rho=1}^{r} \zeta^{(\rho)}_{\gamma;\alpha} \overline{\zeta^{(\rho)}_{\delta;\beta}}$$

(Bochner [2]), where the semi-colon indicates again covariant differentiation with respect to the metric (9.15) thus created. Now, (9.19) implies

$$- R_{\alpha\bar{\beta}} \xi^{\alpha} \xi^{\bar{\beta}} = g^{\gamma\bar{\delta}} \sum_{\rho=1}^{r} \zeta^{(\rho)}_{\gamma;\alpha} \xi^{\alpha} \zeta^{(\rho)}_{\bar{\delta};\bar{\beta}} \xi^{\bar{\beta}}$$

and thus we have

$$(9.20) \qquad R_{\alpha\bar{\beta}} \xi^{\alpha} \xi^{\bar{\beta}} \leq 0$$

with equality holding only if

$$(9.21) \qquad \zeta^{(\rho)}_{\gamma;\alpha} \xi^{\alpha} = 0$$

Hence the following conclusion:

THEOREM 9.6. If a compact Kaehler manifold is such that its metric tensor $g_{\alpha\bar{\beta}}$ can be obtained, in the neighborhood of every point, from an isometric analytic imbedding into a flat Kaehler manifold, then the Ricci curvature is non-positive, and therefore there exists no contravariant analytic vectors and tensors

$$\xi^{\alpha_1 \alpha_2 \ldots \alpha_p}$$

other than those having covariant derivative zero; and, for instance, a non-zero vector ξ^{α} must satisfy all equations (9.21) for all γ and ρ .

Thus, in particular, if a compact complex manifold can be imbedded analytically locally one-one into a complex multi-torus of some dimension, then on the manifold any analytic tensor

$$\xi^{\alpha_1 \alpha_2 \dots \alpha_p}$$

is determined by its value at any one point, and it is a parallel field in the induced metric on the manifold.

It should be noted that the Ricci curvature remains non-positive, if, more generally than in the theorem, the metric tensor $g_{\alpha\bar{\beta}}$ given is such that in a neighborhood of every point, its Ricci curvature is the uniform limit of expressions (9.19); and the theory of automorphic functions of one or several variables is replete with metrics of this kind. Thus the classical hyperbolic metric

$$ds^2 = \frac{dz d\bar{z}}{(1 - z\bar{z})^2}$$

in the unit circle $|z| < 1$ with constant negative curvature is such a limit but it can be proved that it cannot be obtained by imbedding into a finite dimensional flat Kaehler metric directly, and this is also true for the multi-dimensional generalization

$$ds^2 = \frac{\sum_\alpha |dz^\alpha|^2 + \left(\sum_\alpha |z^\alpha|^2 \sum_\beta |dz^\beta|^2 - |\sum_\alpha \bar{z}^\alpha dz^\alpha|^2 \right)}{\left(1 - \sum_\alpha |z^\alpha|^2\right)^2} \tag{9.22}$$

in the unit sphere

$$|z_1|^2 + |z_2|^2 + \dots + |z_n|^2 < 1 \tag{9.23}$$

and for many other spaces of which we only mention the general hyperbolic matrix space which is a generalization of the last one. It arises, if instead of the n variables $z^1, z^2, \dots, z^n$, we take any rectangular matrix

$$Z = \{z^{\lambda\mu}\} \tag{9.24}$$

$\lambda = 1, 2, \dots, m$; $\mu = 1, 2, \dots, n$, of independent complex variables whose total number then is mn . The unit sphere is then to be replaced by the point set which is characterized by the fact that the square matrix

$$\sum_{\lambda=1}^{m} z^{\lambda\rho} \overline{z^{\lambda\sigma}}$$

has all eigenvalues in absolute value less than 1, that is, by the fact that for any complex numbers $\lambda_1, \lambda_2, \ldots, \lambda_n$, we have

$$\sum_{\rho,\sigma=1}^{n}\sum_{\lambda=1}^{m} z^{\lambda\rho}\lambda_\rho\overline{z^{\lambda\sigma}\lambda_\sigma} < \sum_{\rho=1}^{n}|\lambda_\rho|^2$$

and the corresponding line-element is then

$$ds^2 = \mathrm{trace}[(I_m - ZZ^*)^{-1}dZ(I_n - Z^*Z)^{-1}dZ^*]$$

where Z^* is the matrix adjoint to (9.24) and I_m and I_n are the unit matrices of m and n dimensions respectively.

5. SUFFICIENTLY MANY VECTOR OR TENSOR FIELDS

In Theorem 9.6 we have shown that if there are sufficiently many analytic vectors $\zeta_\alpha^{(\rho)}(z)$ which are suitably "independent," and if they satisfy the curl assumption (9.12), then there are no contravariant vector or tensor fields (unless rather special ones), and it was this curl assumption which introduced the Ricci curvature into the context and secured the conclusion as a special case of a previous general theorem.

Now, the fact is, that the curl assumption can be dropped entirely, and the curvature can be eliminated from the context altogether, provided we sharpen the independence assumption to a certain extent. Actually, our new independence assumption is considerably more exacting than the previous one, but not on the face of it, and the different approach has many interesting implications for algebraic geometry anyway.

If ξ_α is covariant and η^α contravariant then $\xi_\alpha\eta^\alpha$ is a scalar, and an analytic scalar on a compact analytic manifold is a constant,

(9.25) $$\xi_\alpha\eta^\alpha = c$$

with no metric whatsoever required. If now r vectors $\xi_\alpha^{(\rho)}$ are given, then we have the system of equations

(9.26) $$\xi_\alpha^{(\rho)}\eta^\alpha = c^\rho \qquad \rho = 1, 2, \ldots, r$$

with certain constants c^ρ, and it is not hard to obtain part (i) of the following theorem.

THEOREM 9.7. (i) If on a compact complex manifold there are given r analytic covariant vector fields $\xi_\alpha^{(\rho)}$, $r \geq n + 1$, and if they have the property that for any system of constants $c^1, c^2, \ldots, c^r$ not all zero, in a neighborhood of some point the rank of the matrix

(9.27) $$\{\xi_1^{(\rho)}, \xi_2^{(\rho)}, \ldots, \xi_n^{(\rho)}, c^{(\rho)}\}_{\rho=1,2,\ldots,r}$$

has its maximum value $n + 1$, then there exists no analytic contravariant vector fields η^α other than zero.

(ii) Furthermore there exists no analytic contravariant tensor field

$$\eta^{\alpha_1\alpha_2\cdots\alpha_p}$$

other than zero, (Bochner [8]).

For the still remaining proof of part (ii) we will apply induction on p . The r contractions

$$\xi_{\alpha_p}^{(\rho)}\eta^{\alpha_1\alpha_2\cdots\alpha_{p-1}\alpha_p}$$

are tensors in $p - 1$ upper indices, and if the theorem is already known for $p - 1$ they must be all zero, and therefore

(9.28) $$\xi_{\alpha_p}^{(\rho)}\eta^{\alpha_1\alpha_2\cdots\alpha_{p-1}\alpha_p} = 0$$

However, our assumption on the rank of (9.27) implies in particular that the matrix $\{\xi_\alpha^{(\rho)}\}$ itself must have rank n , and therefore (9.28) implies

$$\eta^{\alpha_1\alpha_2\cdots\alpha_p} = 0$$

as claimed.

Actually, part (i) of Theorem 9.7 can be generalized from vectors to tensors in the following manner. For any (r, s) consider the mixed tensor

(9.29) $$\xi_{\alpha_1\cdots\alpha_r}^{\beta_1\cdots\beta_s}$$

and then, correspondingly, the mixed tensor

$$\eta^{\alpha_1 \ldots \alpha_r}{}_{\beta_1 \ldots \beta_s} \tag{9.30}$$

of the complementary type (s, r), and the assertion is that for given (r, s) if there are sufficiently many suitably independent tensors of the type (9.29) then there is no tensor of type (9.30). In order to express the necessary assumptions in a simple manner we are introducing the following notation which has been elaborated systematically in another context (Bochner [10]).

In general, the tensor (9.29) has $N = n^{r+s}$ components, and we now denote these components in a fixed ordering by

$$\xi_A \qquad A = 1, 2, \ldots, N$$

Correspondingly, the components of (9.30) are to be denoted by η^A, and the contraction of the tensors is $\xi_A \eta^A$. Now, if both are analytic on a compact manifold, then we must have

$$\xi_A \eta^A = c$$

and the conclusion is as follows:

THEOREM 9.8. If on a compact complex manifold there are given r tensor fields $\xi_A^{(\rho)}$, $r \geq N + 1$, and if for every system of constants

$$(c^1, c^2, \ldots, c^r) \neq (0, 0, \ldots, 0)$$

the rank of the matrix

$$\{\xi_1^{(\rho)}, \xi_2^{(\rho)}, \ldots, \xi_N^{(\rho)}, c^\rho\}_{\rho=1,2,\ldots,r} \tag{9.31}$$

is $N + 1$ at some point, then there exists on the manifold no analytic tensor field η^A of the complementary type.

The assumption in the theorem is the more restrictive the larger the number N is, but this number can be reduced if there are "symmetries" or "anti-symmetries" and more general dependencies between the components of the tensors. Thus, for instance, the following theorem is true in which the number N has been reduced from the original value n^n to 1.

THEOREM 9.9. If on a compact complex manifold

there are two analytic anti-symmetric tensor fields

$$\xi^{(1)}_{\alpha_1\alpha_2\ldots\alpha_n} , \quad \xi^{(2)}_{\alpha_1\alpha_2\ldots\alpha_n}$$

which are linearly independent, then there exists no anti-symmetric

$$\eta^{\alpha_1\alpha_2\ldots\alpha_n}$$

and in particular there cannot then exist n vector fields $\eta^{\alpha}_{(\rho)}$, $\rho = 1, 2, \ldots, n$, whose determinant

$$|\eta^{\alpha}_{(\rho)}|_{\alpha, \rho=1,2,\ldots,n}$$

is $\not\equiv 0$, meaning that there cannot exist on it a complex Lie group of analytic isomorphisms which is transitive.

Finally, one last remark about vectors on a real compact Riemannian manifold. On such one we have $\xi_i\eta^i = c$ if ξ_i is harmonic and η^i is a Killing vector, and the reasoning applies again, and without assumptions on curvature the result follows that if there are sufficiently many suitably independent vectors of one kind then there are no vectors of the other kind. But, we are not stating this result formally, because in the present case the previous theorems involving curvature are by and large considerably better than this theorem would be.

6. EULER-POINCARÉ CHARACTERISTIC

The oldest connection between curvature and Betti numbers is the Gauss-Bonnet theorem. Its recent development falls into directions somewhat different from those pursued in this tract; but there are points of contact nevertheless, and we will quote first one result which falls in the direction of our previous section 4 and for whose proof we refer to Bochner [7].

THEOREM 9.10. If on a compact Kaehler manifold the metric tensor in the neighborhood of every point can be written in the form

$$g_{\alpha\bar{\beta}} = \sum_{\rho=1}^{r} \zeta^{(\rho)}_{\alpha}\overline{\zeta^{(\rho)}_{\beta}} \tag{9.32}$$

with $r \geq n$, where the tensors $\zeta_{\alpha}^{(\rho)}(z)$ have the curl property

$$\frac{\partial \zeta_{\alpha}^{(\rho)}}{\partial z^{\beta}} = \frac{\partial \zeta_{\beta}^{(\rho)}}{\partial z^{\alpha}}$$

and the matrix

$$\{\zeta_{\alpha}^{(\rho)}\}$$

has rank n , then the Euler-Poincaré characteristic has the algebraic sign $(-1)^{n}$ or has value zero.

This conclusion also holds, if more generally in the neighborhood of every point the metric tensor $g_{\alpha\bar{\beta}}$ is a uniform limit, as $s \longrightarrow \infty$, of tensors $g_{\alpha\bar{\beta}}^{(s)}$ each of which has the form stated and if the first and second derivatives of $g_{\alpha\bar{\beta}}^{(\rho)}$ are likewise uniformly convergent towards those of $g_{\alpha\bar{\beta}}$.

Furthermore, if the metric tensor has strictly the form (9.32) for one system of vectors $\zeta_{\alpha}^{(\rho)}$ for the entire manifold, then the characteristic has the value zero if and only if the sums

$$\epsilon^{\alpha_1 \cdots \alpha_n} \epsilon^{\beta_1 \cdots \beta_n} \zeta_{\alpha_1;\beta_1}^{(\rho_1)} \cdots \zeta_{\alpha_n;\beta_n}^{(\rho_n)}$$

are all zero for all $1 \leq \rho_1 \leq r, \ldots, 1 \leq \rho_n \leq r$, the symbols

$$\epsilon^{\alpha_1 \alpha_2 \cdots \alpha_n} , \epsilon^{\beta_1 \beta_2 \cdots \beta_n}$$

being the known Kronecker symbols.

7. NON-COMPACT MANIFOLDS AND BOUNDARY VALUES ZERO

Many of our leading theorems followed from the lemma that if on a compact Riemannian manifold V_n we have $\Delta\phi \geq 0$ everywhere then we have $\phi = c$ and therefore $\Delta\phi = 0$, and the compactness of the manifold was used only for making certain that the given continuous function ϕ attains its maximum.

Assume now that our manifold V_n is arbitrary non-compact, but assume then that the continuous function ϕ has "boundary values zero" in the following sense: To each ϵ there is a compact subset D^{ϵ} of V_n

such that in $V_n - D^\epsilon$ we have $|\phi| < \epsilon$. The function ϕ has then again a maximum and the lemma applies, and we obtain the following conclusion. If the Ricci curvature is positive, and if for a vector field ξ_i with

$$\xi_{i;j} = \xi_{j;i} \quad \text{and} \quad \xi^i{}_{;i} = 0$$

the scalar function $\phi = \xi^i \xi_i$ has "boundary values zero" in the manner stated then we have $\xi_i \equiv 0$. Other theorems which can be so generalized are: Theorems 2.10, 3.1, 3.2, 3.3, 3.5, 7.4, 7.5, 7.6, 7.7, 7.8, 7.9, 7.13, 7.14, 7.15.

Note also that syllogistically this is a true generalization, because if V_n is compact, and ϕ is arbitrary continuous, we can take for D^ϵ the entire manifold V_n itself and on the (empty) set $V_n - D^\epsilon$ we then have $|\phi| < \epsilon$, as required.

8. ALMOST-AUTOMORPHIC VECTOR AND TENSOR FIELDS

A less trivial generalization arises in the following manner. Assume V_n compact, but introduce its universal covering space $\tilde{V}_n$, even if it is non-compact, and make no restriction on the nature of the fundamental group $\Gamma = \tilde{V}_n/V_n$. If the elements of Γ are

$$\gamma_0, \gamma_1, \gamma_2, \cdots$$

then each γ_r defines a homeomorphism of $\tilde{V}_n$. With each point $\tilde{P}$ of $\tilde{V}_n$ we associate the sequence of points

$$(9.33) \qquad \tilde{P}_r = \gamma_r(\tilde{P}) \qquad r = 0, 1, 2, \ldots$$

and any two points of this sequence are "equivalent." The original manifold V_n can be identified with the space of sequences

$$(9.34) \qquad P = \{\gamma_0(\tilde{P}), \gamma_1(\tilde{P}), \gamma_2(\tilde{P}), \cdots\}$$

in a suitable manner, and on the other hand there is a compact subset R in $\tilde{V}_n$ such that to any point in $\tilde{V}_n$ there is a point equivalent to it. For any given sequence (9.34) we may say that each $\tilde{P}_r$ "covers" P or "lies over" P, and conversely that P is a "projection" of $\tilde{P}_r$.

If we are given any structure on V_n, differentiable or analytic, then there is a structure on $\tilde{V}_n$ of which the given one is a projection, and this structure of $\tilde{V}_n$ is "periodic" (or "automorphic") in the sense that if U is a coordinate neighborhood of $\tilde{V}_n$ and γ_r is an element of

Γ then $\gamma_r(U)$ is again such a coordinate neighborhood.

We say that a function on $\tilde{V}_n$, scalar or tensor, is periodic (or automorphic) if we have $\phi(\gamma_r\tilde{P}) = \phi(\tilde{P})$ for all r, and $\xi_i(\gamma_r\tilde{P}) = \xi_i(\tilde{P})$ for a vector, and in the same manner for any tensor. Any periodic function on $\tilde{V}_n$ gives rise to a function on V_n itself (its "projection") by putting $\phi(P) = \phi(\tilde{P})$, and conversely any function $\phi(P)$ on V_n gives rise to a periodic function on $\tilde{V}_n$ by putting $\phi(\gamma_r\tilde{P}) = \phi(P)$. In particular if we are given a metric tensor g_{ij} on V_n then it has a periodic extension onto $\tilde{V}_n$, and we will denote it again by g_{ij}.

Now, we call a continuous function $\phi(\tilde{P})$ on $\tilde{V}_n$ "almost periodic" (or "almost automorphic," in either case "relative to the given group Γ") if every infinite sequence of elements $\{\gamma_s\}$ contains an infinite subsequence such that the sequence of functions

$$\{\phi(\gamma_r\tilde{P})\} \qquad r = 1, 2, \ldots,$$

is convergent uniformly on the entire space V_n. The definition also applies to vectors and tensors, the uniformity of convergence being relative to the uniform structure of the space, and the best way of expressing this uniformity is to utilize the tensor g_{ij}, assumed periodic, in the following manner: given, say, a vector $\xi_i(\tilde{P})$ then the sequence of "translated" vectors $\xi_i(\gamma_r\tilde{P})$ converges uniformly towards a limiting vector $\xi_i^*(\tilde{P})$ if the square length

$$g^{ij}(\tilde{P})[\xi_i(\gamma_r\tilde{P}) - \xi_i(\tilde{P})][\xi_j(\gamma_r\tilde{P}) - \xi_j(\tilde{P})]$$

converges to 0, as $r \longrightarrow \infty$, uniformly in V_n, and similarly for tensors.

Now, almost periodic functions and tensors have the following properties. First of all, due to the compactness of the set R previously introduced, any continuous periodic function is almost-periodic, and any almost periodic function is bounded. A constant function is of course almost periodic. The sum and the product of two almost periodic functions, scalar or tensor, is almost periodic, and the contractions of an almost periodic tensor is again almost periodic, and finally there is the following property which will be all-decisive in our argument. If a function $\phi(\tilde{P})$, scalar or tensor, is almost periodic, and if for a sequence of elements $\{\gamma_r\}$ the sequence $\phi(\gamma_r\tilde{P})$ converges uniformly, and if we denote the limit function by $\phi^*(\tilde{P})$, and if we denote by γ_r^{-1} the group element inverse to γ_r, then the sequence of function $\phi^*(\gamma_r^{-1}\tilde{P})$ converges back to the original function $\phi(\tilde{P})$.

Now, Theorem 2.3 can be given the following generalization.

THEOREM 9.11. If a function $\phi \in C^2$ on $\tilde{V}_n$ is such that the function ϕ and its derivatives $\phi_{;i}$, $\phi_{;i;j}$ are all almost periodic, and if we have $\Delta\phi \geq 0$ on $\tilde{V}_n$, then we have $\phi = c$, and thus again $\Delta\phi = 0$.

PROOF. Being almost periodic, the function is bounded and if $M = \sup \phi(P)$ then there is a sequence of points $\tilde{Q}_r$ such that $\phi(\tilde{Q}_r) \longrightarrow M$. Now there is an element γ_r^{-1} such that $\tilde{P}_r = \gamma_r^{-1}\tilde{Q}_r$ lies in R; and R being compact, and our function being almost periodic we may assume (after taking an infinite subsequence perhaps) that the following limit relations hold. First of all the sequence of points $\tilde{P}_r$ in R converges to a limit point in R which we will denote by $\tilde{P}^*$, and secondly all functions $\phi(\gamma_r\tilde{P})$, $\phi(\gamma_r\tilde{P})_{;i}$, $\phi(\gamma_r\tilde{P})_{;i;j}$ are uniformly convergent, and we will denote the limit of the first by $\phi^*(\tilde{P})$. Now, due to the uniform convergence of all functions we first of all have

$$\Delta\phi^* \geq 0 \tag{9.35}$$

and secondly since

$$\phi(\gamma_r\tilde{P}) = \phi(\tilde{Q}_r) \longrightarrow M$$

it follows from the uniform convergence that for the limiting function ϕ^* we have $\phi^*(\tilde{P}^*) = M$, and thus $\phi^*(\tilde{P}^*)$ assumes its maximum at a point. But from this and (9.35) it follows now that $\phi^* =$ constant. However, and this last step is decisive, from the limiting function ϕ^* we can reobtain the original function $\phi(\tilde{P})$ as the limit of $\phi^*(\gamma_r^{-1}\tilde{P})$, but since ϕ^* has just proven to be a constant, the original function ϕ must have been a constant, as claimed.

Once Theorem 9.11 has been established many theorems can be generalized, and we only formulate one.

THEOREM 9.12. If ξ_i is defined on $\tilde{V}_n$ and satisfies

$$\xi_{i;j} = \xi_{j;i} \quad \text{and} \quad \xi^i{}_{;i} = 0 \tag{9.36}$$

and if ξ_i, $\xi_{i;j}$ and $\xi_{i;j;k}$ are all continuous and almost periodic, then for positive Ricci curvature, ξ_i must be identically zero.

We are of course entitled to calling a vector with property (9.36) a harmonic (almost periodic) vector on $\widetilde{V}_n$, but there is no immediate interpretation of this in terms of cohomology known to us, and to the contrary, perhaps the best way of <u>defining</u> an almost periodic co-cycle would be to start with the definition (9.36) of harmonic vector and then make the topological definition fit the differential geometric data.

Furthermore, in the above Theorems 9.11 and 9.12, the metric tensor itself was assumed periodic. But we might have allowed this tensor itself to be almost periodic only, and the conclusion would have been the same.

BIBLIOGRAPHY

BOCHNER, S.

[1] "Remarks on the theorem of Green," Duke Math. Journal 3 (1937), 334-338.

[2] "Vector fields and Ricci curvature," Bull. of the Amer. Math. Soc. 52 (1946), 776-797.

[3] "Curvature in Hermitian manifolds," Bull. of the Amer. Math. Soc. 53 (1947), 179-195.

[4] "On compact complex manifolds," Journal of Indian Math. Soc. 11 (1947), 1-21.

[5] "Curvature and Betti numbers," Ann. of Math. 49 (1948), 379-390.

[6] "Curvature and Betti numbers, II," Ann. of Math. 50 (1949), 77-93.

[7] "Euler-Poincaré characteristic for locally homogeneous and complex spaces," Ann. of Math. 51 (1950), 241-261.

[8] "Vector fields on complex and real manifolds," Ann. of Math. 52 (1950), 642-649.

[9] "Complex spaces with transitive commutative groups of transformations," Proc. Nat. Acad. Sci. U.S.A. 37 (1951), 356-359.

[10] "A new viewpoint in differential geometry," Canadian Journal of Math. 3 (1951), 460-470.

[11] "Tensorfields and Ricci curvature in Hermitian metric," Proc. Nat. Acad. Sci. U.S.A. 37 (1951), 704-706.

[12] "Laplace operator on manifolds," Proceedings of the International Congress of Mathematicians, Vol. II (1952), 189-201.

BOCHNER, S., YANO, K.

[1] "Tensor-fields in non-symmetric connections," Ann. of Math. 56 (1952), 504-519.

CHERN, S. S.

[1] "Characteristic classes of Hermitian manifolds," Ann. of Math. 47 (1946), 85-121.

de RHAM, G.

[1] "Remarque au sujet de la théorie des formes différentielles harmoniques," Annales de l'Université de Grenoble 23 (1947-48), 55-56.

de RHAM, G., KODAIRA, K.

[1] "The harmonic integrals," Lectures delivered at the Institute for Advanced Study (1950).

ECKMANN, B., GUGGENHEIMER, H.

[1] "Formes différentielles et métrique hermitienne sans torsion, I. Structure complexe, formes pures," C. R. Acad. Paris 229 (1949), 464-466.

[2] "Formes différentielles et métrique hermitienne sans torsion, II. Formes de classe k; formes analytiques," C. R. Acad. Sci. Paris 229 (1949), 489-491.

[3] "Sur les variétés closes à métrique hermitienne sans torsion," C. R. Acad. Sci. Paris 229 (1949), 503-505.

[4] "Quelques propriétés globales des variétés hermitiennes," C. R. Acad. Sci. Paris, 229 (1949), 577-579.

EISENHART, L. P.

[1] Riemannian Geometry (1926).

[2] Continuous Groups of Transformations (1933).

FUBINI, G.

[1] Teoria dei Gruppi Discontinui (1908).

GARABEDIAN, P. R., SPENCER, D. C.

[1] "A complex tensor calculus for Kaehler manifold," Technical Report 17, Stanford University (1951).

GUGGENHEIMER, H.

[1] "A note on curvature and Betti numbers," Proc. Amer. Math. Soc. 2 (1951), 867-870.

[2] "Ueber komplex-analytische Mannigfaltigkeiten mit Kaehlerscher Metrik," Comment. Math. Helv. 25 (1951), 257-297.

HODGE, W. V. D.

[1] The Theory and Applications of Harmonic Integrals, Cambridge Univ. Press (1952).

HOPF, E.

[1] "Elementare Bemerkungen über die Lösungen partieller Differentialgleichungen zweiter Ordnung vom elliptischen Typus," Sitzungsber. Preuss. Akad. Wiss. 19 (1927), 147-152.

HOPF, H.

[1] "Zum Clifford-Kleinschen Raumproblem," Math. Ann. 95 (1925), 313-339.

KAEHLER, E.

[1] "Ueber eine bemerkenswerte Hermitische Metrik," Abh. Math. Sem. Hamburgischen Univ. 9 (1933), 173-186.

LICHNÉROWICZ, A.

[1] "Courbure et nombres de Betti d'une variété riemannienne compacte," C. R. Acad. Sci. Paris 226 (1948), 1678-1680.

[2] "Dérivation covariante et nombres de Betti," C. R. Acad. Sci. Paris 230 (1950), 1248-1250.

[3] "Théorèmes de réductibilité des variétés kaehleriennes et applications," C. R. Acad. Sci. Paris 231 (1950), 1280-1282.

[4] "Sur les variétés riemanniennes admettant une forme quadratique à dérivée covariante nulle," C. R. Acad. Sci. Paris 231 (1950), 1413-1415.

[5] "Formes à dérivée covariante nulle sur une variété riemannienne," C. R. Acad. Sci. Paris 232 (1951), 146-147.

[6] "Sur les variétés riemanniennes admettant une forme à dérivée covariante nulle," C. R. Acad. Sci. Paris 232 (1951), 677-679.

[7] "Sur les formes harmoniques des variétés riemanniennes localement réductibles," C. R. Acad. Sci. Paris 232 (1951), 1634-1636.

[8] "Généralization de la géométrie kaehlerienne globale," Colloque de géométrie différentielle (1951), 99-122.

[9] "Courbure, nombres de Betti, et espaces symétriques," Proceedings of the International Congress of Mathematicians, Vol. II (1952), 216-223.

LICHNÉROWICZ, A., WALKER, A. G.

[1] "Sur les espaces riemanniennes harmoniques de type hyperbolique normal," C. R. Acad. Sci. Paris 221 (1945), 394-396.

MOGI, I.

[1] "On harmonic field in Riemannian manifold," Kōdai Math. Seminar Reports 2 (1950), Nos. 4 and 5, 61-66.

MYERS, S. B.

[1] "Riemannian manifolds with positive mean curvature," Duke Math. Journal 8 (1941), 401-404.

RAUCH, H. E.

[1] "A contribution to differential geometry in the large," Ann. of Math. 54 (1951), 38-55.

RUSE, H. S.

[1] "On simply harmonic spaces," Journal of the London Math. Soc. 21 (1946), 243-247.

THOMAS, T. Y.

[1] "Some applications of Green's theorem for compact Riemann spaces," Tohoku Math. Journal 46 (1940), 261-266.

TOMONAGA, Y.

[1] "On Betti numbers of Riemannian spaces," Journal of the Math. Soc. of Japan 2 (1950), 93-104.

VEBLEN, O., WHITEHEAD, J. H. C.

[1] Foundations of Differential Geometry, Cambridge Univ. Press (1932).

WALKER, A. G.

[1] "On completely harmonic spaces," Journal of the London Math. Soc. 20 (1945), 159-163.

WEIL, A.

[1] "Sur la théorie des formes différentielles attachées à une variété analytique complexe," Comment. Math. Helve. 20 (1947), 110-116.

YANO, K.

[1] "Concircular geometry, I, II, III, IV, V," Proc. Imp. Acad. Japan 16 (1940), 195-200; 345-360; 442-448; 505-511; 18 (1942), 446-451.

[2] Groups of Transformations in Generalized Spaces, Tokyo (1949).

[3] "On harmonic and Killing vector fields," Ann. of Math. 55 (1952), 38-45.

[4] "Some remarks on tensor fields and curvature," Ann. of Math. 55 (1952), 328-347.

[5] "On Killing vector fields in a Kaehlerian space," Journal of the Math. Soc. of Japan 5 (1953), 6-12.

Lightning Source UK Ltd.
Milton Keynes UK
UKHW052202220922
409299UK00001B/23